> LEONARDO'S LAPTOP

> LEONARDO'S LAPTOP

HUMAN NEEDS AND THE NEW COMPUTING TECHNOLOGIES

Ben Shneiderman

The MIT Press   Cambridge, Massachusetts   London, England

First MIT Press paperback edition, 2003

©2002 Massachusetts Institute of Technology

This book was set in Garamond Three by Achorn Graphic Services Inc.

Printed and bound in the United States of America.

Library of Congress Cataloging-in-Publication Data

Shneiderman, Ben.

Leonardo's laptop : human needs and the new computing technologies / Ben Shneiderman.

p. cm.

Includes bibliographical references and index.

ISBN 0-262-19476-7 (hc. : alk. paper), 0-262-69299-6 (pb)

1. Electronic data processing. 2. Human-computer interaction. 3. Technological
forecasting. I. Title.

QA76 .S5145 2002

004—dc21

2001056277

1 0 9 8 7 6 5

For Jenny, Sara, Anna

CONTENTS

Computing today is about what computers can do; the new computing will be about what people can do. As users of contemporary technology, we are often angry and frustrated because computers are not in harmony with our needs and abilities. We feel powerless and do not see a role for the individual in the process of technological innovation. This book is primarily designed to raise users expectations of what they should get from technology and to empower both users and developers to invent computers that enhance our lives and our world.

We can accelerate the movement toward the "new computing" by adopting Leonardo da Vinci as an inspirational muse. His integrative spirit, combining science with art and engineering with aesthetics, can help us envision more successful and satisfying experiences with information and communications technologies.

With our new sense of empowerment, we can challenge technology developers to more diligently focus on user needs so that they can produce more effective technologies. Such improvements can come quickly to learning, commerce, healthcare, and government. Small innovations such as jewel-like medical sensors and fingertip computing, will be coupled with large innovations such as secure and immediate worldwide access to medical information and million-person communities that bridge the digital divide and enable consensus seeking and conflict resolution.

ACKNOWLEDGMENTS

So many people helped me to see the future and inspired me. Several of them helped by commenting on draft chapters: Ben Bederson, Alison Druin, Amy Friedlander, Christopher Fry, Jean Gasen, Ruth Guyer, Brian Kahin, Bill Killam, Charles Kreitzberg, Jonathan Lazar, Nancy Leveson, Peter Levine, Henry Lieberman, Richard Mushlin, Catherine Plaisant, Ron Rice, Ariel Sarid, George Schneider, Norman Schneiderman, Barbara Tversky, and Ron Weissman. I am especially appreciative of the people who read the full manuscript at various stages in its development: Harry Hochheiser, Bill Kules, Kent Norman, Stephan Parker, Arkady Pogostkin, Jenny Preece, Anne Rose, Helen Sarid, and the anonymous reviewers. My students over the years, graduates and undergraduates, have been more than a testing ground for these ideas, they are the continuing motivation for my development. I hope they will put these ideas to work and then refine them still further. A continuing inspiration and source of wisdom is my partner and wife, Jenny Preece.

College Park, Maryland

> LEONARDO'S LAPTOP

Lady with an Ermine. From license-free "Leonardo da Vinci: Selected Works," Planet Art.

1

> INSPIRATION FOR THE NEW
COMPUTING

The old computing was about what computers could do; the new computing is about what users can do. Successful technologies are those that are in harmony with users' needs. They must support relationships and activities that enrich the users' experiences.

Information and communication technologies are most appreciated when users experience a sense of security, mastery, and accomplishment. Then these technologies enable users to relax, enjoy, and explore.

> *Imagine that after a sunrise climb, you reach the summit. You open up your phonecam and send a panoramic view to your grandparents, parents and friends. They hear the sound of birds, smell of mountain air, feel the coolness of wind, and experience your feeling of success. They can hear each other cheering, and point at the birds or click on other peaks to find out more. They remember how, on your last climb, a rockslide brought you to an emergency room unconscious. On that occasion, fortunately, your World Wide Med records guided the physician to care for you. She was able to review your medical history in her local language, helping her to prescribe the right treatment. Today's climb has a happier outcome, which restores everyone's confidence.*

The challenge for technology developers is to more deeply understand what you, the user, want. Then they can respond to this challenge by creating products that are more useful and satisfying to more people.

The time is right for the high-tech world to attend more closely to the needs of humanity. Many people are not satisfied with current technologies that make them feel incompetent or unsuccessful. Others can't benefit from technology because of high cost, unnecessary complexity, and lack of relevance to their needs. The new computing must be innovative, and it must focus on raising user satisfaction, broadening participation, and supporting meaningful accomplishment. All this is becoming possible today because the underlying technology is at hand and researchers are finding better techniques to discover what people want.

Computing technology is at a crossroads. The British scientist C. P. Snow wrote about the troubling split between science and art in his lecture on "Two Cultures." He identified a modern dilemma that should be resolved with a second Renaissance, or maybe Renaissance 2.0. This modern Renaissance would unify thinking about technology by promoting multidisciplinary education and a sympathy for diversity. It would emphasize collaborations that enrich us with fresh perspectives and foster partnerships that enable us to create more freely.

However, linking the high-tech world more closely to the needs of people still requires some new forms of thinking. The Renaissance integration of disciplines that Leonardo da Vinci exemplified could guide us in repairing the split in our modern world. Leonardo integrated engineering with human values. He blended science and art to produce graceful drawings of human anatomy, flowing water, and innovative machines. Leonardo-like thinking could help users and technology developers to envision the next generation of information and communication technologies.

The creative genius of Leonardo da Vinci (1452–1519) has inspired technologists, scientists, and artists for more than half a millennium. His Renaissance integration of engineering with human values could be the path to appealing artifacts and provocative dreams.

What I like about Leonardo is that he was more than just a Renaissance geek. His playful side flourished in performing on the lyre and staging musical events. He even fabricated theatrical sets, complete with dancing lion puppets. This combination of skills delights us even today and can suggest future toys and entertainment.

Leonardo appreciated the importance of ambitious visions. His massive bronze horse to honor the father of Ludovico Sforza of Milan was intended to astound viewers with its size, its accurate anatomy, and a graceful ferocity that celebrated courage and strength. However, casting a 24-foot-high statue was beyond the capabilities of fifteenth-century metalworkers. Leonardo, undaunted, planned to make the casts in components. He built a plaster model to impress onlookers and promote the project, but politics interfered and in 1499 the invading French archers destroyed it—merely for target practice.[1] What are the dreams we have for ambitious and inspirational technology projects?

We still admire his skill in producing treasured artworks. The dramatic fresco of the *Last Supper* pleases us with a composition of architectural space that uses perspective to frame the detailed portraits of four groups of three apostles, each with compelling emotional expressions. Leonardo mastered the artistic methods of light and shadow, the mathematical elements of symmetric alignments, and the iconic power of downturned hands and upraised arms. By comparison the iconic language of graphic user interfaces and the World Wide Web seems impoverished. Where are the graphics geniuses and the Web-designer Leonardos whose work stirs and thrills us?

During his lifetime, Leonardo was famous for his public art pieces and his portraits, although we know him also for his inventions of helicopters, submarines, and other mechanical devices. His engineering innovations were often a secret, locked in his notebooks with his medical drawings, insights about geology, optics, hydraulics, and much more. Over the centuries many people have been struck by Leonardo's integration of art with science, and aesthetics with engineering (Kemp 2000).

His notebook pages demonstrate the benefits of integrating graphics and text, and his analyses testify to the power of combining visual and analytic thinking. And now again, five hundred and fifty years after his birth, this combination of skills inspires us—this time to envision information and communication technologies that are in harmony with human needs. In this book, I propose Leonardo as an inspirational muse for the *new computing*.

LEONARDO'S HUMBLE START

Uniquely, he was able to see science from the perspective of an artist, to visualize art with the mindset of a scientist; and architecture with the mindset of the artist-scientist. If there is one simple defining skill that distinguishes Leonardo, it is this most useful of talents.
—Michael White, *Leonardo: The First Scientist* (2000), 125

Leonardo began life humbly on April 15, 1452, as the illegitimate son of Ser Piero, a notary in the modest town of Vinci in Italy's fertile Tuscany. Early on, Leonardo impressed his teachers with his rapid learning in math, music, singing, and drawing. When Ser Piero took some of Leonardo's drawing to the great artist Andrea del Verrocchio, Leonardo was invited to apprentice in Verrocchio's workshop.

Giorgio Vasari's (1511–1574) flattering biography of Leonardo, first published around 1550, rhapsodizes about young Leonardo: "possessing so divine and wondrous an intelligence, and being a very fine geometrician, Leonardo not only worked in sculpture but in architecture. . . . He drew so carefully and so well on paper that no one has ever matched the delicacy of his style" (Vasari 1998). In a famous story, Verrocchio comments that Leonardo's completion of an angelic figure was so masterful that Verrocchio considered giving up painting. But Leonardo responded graciously that it was the greatest compliment to the master that the student should exceed the master's ability. The model of teamwork in Verrocchio's studio probably influenced Leonardo to build a community around him that later in life included the noted mathematician Luca Pacioli as well as devoted younger artists such as Andrea Salai and Francesco Melzi.

Leonardo's remarkable capacity for observation was supported by a purposeful focus that led him to ask the right questions. His sharp eyes and mind enabled him to make discoveries and innovations in fields as diverse as medicine, aeronautical engineering, and geology. He was the first to accurately draw and recognize the role of curved spines in humans (figure 1.1), and shocked many with his drawing of a fetus inside a womb (figure 1.2). Leonardo's keen observation of birds led him to make sketches of a parachute and a crude airplane that were four hundred years ahead of his time. His integrative spirit was not unusual in Renaissance Italy, where a conscious blend of scientific invention with aesthetics was common. Logic and art were partners; mathematics and music were collaborators.

But beyond integrating disciplines, Leonardo had a distinctly inquisitive mind and capacity for independent thinking that led him to go further than his contemporaries in many topics. For example, he considered why seashells were found in the Tuscan hills. Contemporary wisdom said that the seashells were washed up into the mountains during the Biblical flood. However, Leonardo noticed seashells at many sedimentary levels and correctly guessed that the Tuscan hills had once been under the ocean. This is accepted twentieth-century science, but it was heresy in the fifteenth century, when church doctrine was still founded on the unchanging nature of the earth. Challenging these deep beliefs took an independent mind and a courageous spirit. Galileo (1564–1642) suffered terribly for merely raising the possibility that the earth might revolve around the sun, a possibility first raised seriously by Copernicus (1473–1543).

The same skills of observation and systematic inquiry empowered Leonardo to draw and paint remarkable images (Clark 1939). He would walk the streets of Florence and return at night to sketch twenty accurate and sympathetic portraits of the peasants and elderly citizens he had seen. His paintings depicting *Mona Lisa* (figure 1.3) and *Ginevra de' Benci* (figure 1.4) fascinate viewers because the compelling faces reveal subtle emotions that invite lengthy contemplation. Leonardo's portraits can be

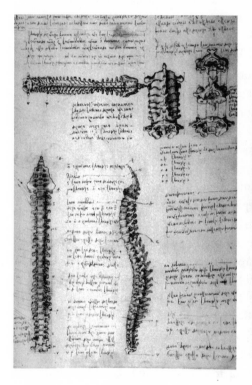
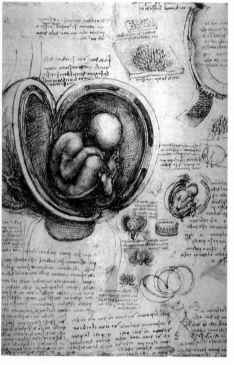

1.1   Leonardo's sketch of the human spine. From license-free "Leonardo da Vinci: Selected Works," Planet Art.

1.2   Leonardo's sketch of a fetus in the womb. From license-free "Leonardo da Vinci: Selected Works," Planet Art.

1.3   Leonardo's portrait *Mona Lisa.* From license-free "Leonardo da Vinci: Selected Works," Planet Art.

1.4   Leonardo's portrait *Ginevra de' Benci.* Ailsa Mellon Bruce Fund, Photograph © 2001 Board of Trustees, National Gallery of Art, Washington, D.C.

interpreted as smiling or smirking, contented or contemptuous. The precise facial details are complemented by the careful choices of background plants, such as the juniper tree, which is *ginevra* in Italian. The orderly compositions in his paintings and frescoes guide the viewer's eyes, demonstrating Leonardo's mastery of architecture and detail. If you visit the Louvre in Paris, you can see the elaborate installation honoring *Mona Lisa*, or if you travel to the National Gallery of Art in Washington, D.C., you can join the crowds in front of Leonardo's portrait *Ginevra de' Benci*.[2]

I would have loved to have seen Leonardo at work. He was an endless doodler, sketcher, and dreamer who tucked several notebooks of varying sizes into his waist belt to record his thoughts. He had small notepads, assorted notebooks, and large folders with fine parchment folios. Scholars estimate that he filled 13,500 pages, of which less than 5,000 survive. His sketches for a submarine and for a helicopter reinforce the characterization of Leonardo as an innovator who was far ahead of the available technology (Wallace 1968).

Leonardo was also a fine self-promoter, exemplified in his consultant's pitch to Ludovico Sforza in Milan, offering to help design war machines, defensive walls, and the massive bronze horse. Leonardo's letter to Sforza shows his capacity to make a convincing business presentation, but in the main Leonardo was devoted to his scientific pursuits, filling his notebooks with observations and speculations. In Sforza's court and later as he traveled, Leonardo kept a household entourage of varied characters. Vasari, always ready to make a compliment, wrote, "His generosity was so great that he sheltered and fed all his friends, rich and poor alike, provided they possessed talent and ability." His late delivery of projects was legendary and his unfulfilled promises made him an easy target for critics. Still, Leonardo was revered in his day and remains a muse who can inspire creative endeavors (Gelb 1998).

During his last years, Leonardo was the honored guest of the French King Francois I, at Amboise. Although he lived in royal surroundings, when Leonardo died at the advanced age of 67, his will contained an unusual request. He wanted to express his lifelong sympathy for the poor and be honored by a funeral procession that included sixty peasants carrying torches.

Leonardo would have been amused that in 1994, Bill and Melinda Gates bought 72 pages of Leonardo's writings, the Codex Leicester, for $30.8 million, and arranged for a well-financed exhibition tour to leading museums and an informative CD-ROM (Corbis 1997).

ENVISIONING THE NEW COMPUTING

The models of Leonardo's inventions in Milan's science museum provoke me to wonder, if Leonardo were alive today, how he would use a laptop and what kind of novel computers he would design. Would Leonardo be employed by Apple to "Think Different" or by Intel and Microsoft to give Windows a Renaissance 2.0 look and feel? Certainly, Leonardo's visual thinking would be important in shaping modern computing environments.

Inspired by Leonardo's penchant for portable notebooks, and larger sketchbooks, and by his frescoes, we as users and technology developers might imagine the need for a comprehensive line of computers from small but elegant wearable devices to ornate desktop machines and impressive wall-sized models. Keeping in the spirit of Leonardo, each new computing model would be delightfully entertaining and com-pellingly useful. A modern Leonardo of software might be inspired to pursue projects such as a precise 3D medical simulations with tactile feedback that lets you crawl through the human body, a complete environmental model of the world to study global change, and a building-sized FrescoMaker drawing package.

The medical simulation would show precise details and allow you to explore down to the level of each muscle cell and nerve synapse (Nuland 2000). You would be able to see each cell functioning, watch genetic processes, and find new relationships. The environmental model would allow you to try a thousand alternative management policies in an hour and communicate them easily to colleagues and decision-makers. The modern FrescoMaker would allow easy reworking of images until every dangling curl of hair was just right, even on a thousand-foot high building facade.

The new computing technologies would include wall-sized displays, palmtop appliances, and tiny jewel-like medical sensors and fingertip computers that change your sensory experiences and ways of thinking. Your understanding of the world would change when you watch an HIV virus invading a healthy cell or a genetic drug stopping a breast cancer. Your health would improve when tiny sensors assure you that the tasty bite of raspberry ice cream has low enough cholesterol to suit your diet.

New computing will immerse you in dramatic projected experiences or it will become invisible, as the technology is embedded in common devices or inserted under your skin. Mobility and ubiquity will become accepted and expected. New computing will enable you to gather the names and e-mail addresses of everyone in a room who gives their permission and to send them all copies of your slides or home page. The movement from independent work to collaboration with distant colleagues will be seamless.

When you plan a trip, you can select itineraries based on your complete travel history with records of your preferences for cities and natural locations balanced with profiles of new museums, scenic parks, or tranquil beach resorts. You can choose historic or natural sites based on comments from people you trust and interview local guides whom you hire for their colorful personality or botanical knowledge. Your planning will create travel experiences that are more emotionally intense and memorable.

When you travel—even while en route supersonically at 40,000 feet or at your remote destination—the new computing will enable you to record and share your experiences with family or colleagues. They'll be able to see what you see, hear what you hear, smell what you smell, and experience your excitement. When you point your palmtop digital guides at the Alamo Monument or the Suez Canal, you'll get a historical, political, or geological summary. Then you can read comments from previous visitors, look at nineteenth century photos, or leave a record or your impressions for others.

When you point your IdentiCam at a bright yellow flower, its name and description will appear. When you point at a red, white, and black striped snake you'll get a warning that "coral snakes are poisonous." The record of your journeys will be preserved using automatic TravelTemplates that combine your photos with professional ones. When you get home, you'll be able to reexamine your climb up Mt. Kilimanjaro or reminisce about your working side-by-side with a Japanese pottery master.

You will have greater choices when you follow sports teams, take up hobbies, and indulge in Web-based entertainment. You won't have to limit yourself to local teams but could follow playoffs anywhere in the world, replay historic tournaments, and simulate games with players selected from any time in history. Families will create detailed multimedia histories, vividly relive weddings, and reenact key events in their ancestors' lives. You won't be able to go back in time, but you will have an intense appreciation of who your ancestors were and how they lived their lives. You'll share these multimedia histories with family and friends using open directories that only your family members can access.

And beyond information and communication, the new computing will emphasize innovation, or maybe e-novation. Computers are tools for doing and making. They fit comfortably with the Renaissance definition of *homo faber,* man the maker. The new computing software that supports innovation will provide exemplars of excellence for you to build on, templates for getting started, and processes for guiding your creative experiences. Even as a novice you'll be able to perform better than today's experts.

THE OLD COMPUTING GIVES BIRTH TO THE NEW COMPUTING

Looking to the past is often a good way to see to the future. During the early development of computers, technology promoters were in the driver's seat. Their destinations were large-scale engineering projects for military or industrial purposes. These founders of the old computing overcame technological limitations to build impressive projects and then turned to producing tools for themselves, giving little thought to the needs of other users.

By the 1980s, with the advent of the personal computer, the steering wheel of innovation was taken over by those who recognized the importance of considering diverse user needs. These spirited innovators came up with the hot products that opened the doors to a wide range of users: graphical user interfaces (GUIs), the World Wide Web, online communities, instant messaging, information visualization, and e-commerce. This shift has accelerated in recent years, and future breakthroughs are likely to come more often from those who put users first.

Of course, we still need good work from professionals of the old computing to create faster processors, larger databases, and more reliable networks, but I believe that the significant future advances will emanate more frequently from thinkers who are in tune with the new computing. They are more likely to recognize and respond to broad markets in which tools to empower users include collaborative experiences, entertainment, and aesthetics.

Progress is already being made in improving users' experience of computing, but too many people still find computers to be frustrating. This book is designed to raise your expectations of what you get from information and communication technologies. It presents a vision of truly helpful technologies in harmony with human needs. You have needs to communicate with friends, organize family vacations, or find information about your health problems. You want to collaborate with professional colleagues, participate in your local, national, or international communities, and find the best deal for your next car. You should be able to do all these things and much more in a hassle-free and confident way. Your attention should be on your goal, not on the technology you are using to accomplish it.

But too often, the old computing designs produce confusion and frustration. Too often they have incomprehensible terminology, poor online assistance, and nasty failures. Too often the complexity of networks, the layers of applications, and the fragility of software result in untimely crashes and unhappy users. These experiences generate anxiety about computers, resistance to using technology, and fear of losing control.

The challenge for new computing developers is to understand what you, the user, want and to help you get it. Developers can then design information and communication technologies that enable you to achieve your goals rapidly and gracefully in an atmosphere of trust and responsibility. You should be able to trust the information sources that you consult, the deals you are offered, and the privacy you are promised. You should be able to take responsibility for your decisions and for your communications to others. The underlying systems should provide an infrastructure that generates a user experience of reliability and security so that you can concentrate on your work and relationships. This transformation is proceeding in leading research centers and progressive companies, but it does meet resistance. In order to encourage the spread of new computing ideas, it will be helpful to understand and make explicit the underlying changes of attitudes.

The first transformation from the old to the new computing is the shift in what users value. Users of the old computing proudly talked about their gigabytes and megahertz, but users of the new computing brag about how many e-mails they sent, how many bids they made in online auctions, and how many discussion groups they posted to. The old computing was about mastering technology; the new computing is about supporting human relationships. The old computing was about formulating query commands for databases; the new computing is about participating in knowledge communities. Teachers no longer cover the subject; they guide learners to discover it. Salespeople no longer sell products; they form customer relationships.

The second transformation to the new computing is the shift from machine-centered automation to user-centered services and tools. Instead of the machine doing the job, the goal is to enable you to do a better job. Automated medical diagnosis programs that do what doctors do have faded as a strong research topic; however, rapid access to extensive medical lab tests plus patient records for physicians are expected, and online medical support groups for patients are thriving. Robots to clean your home are still a playful fantasy, but music downloads and Web-based family photo albums are booming. Natural language dialogues with computerized therapists have nearly vanished, but search engines that enable users to specify their information needs are flourishing. The next generation of computers will bring even more powerful tools to enable you to be more creative and then disseminate your work online. This Copernican shift is bringing concerns about users from the periphery to the center. The emerging focus is on what users want to do in their lives.

As technology developers acknowledge these two transformations, they will more easily understand the new goals. Short-term benefits will emerge in already ascendant applications such as e-learning, e-business, e-healthcare, and e-government services. Longer-term innovations will appear in new forms of employment, interactive entertainment, decentralized political organizations, and empathic online communities.

To spark our imagination about the new computing, let's go further and explore how an extraordinarily creative historical figure, like Leonardo, might reflect on computers. Wouldn't he put people at the center and think about how to apply technology for their benefit? Leonardo wrote boldly, "Work must commence with the conception of man" (White 2000, 166), and he characterized the "four universal states of man" as "mirth, weeping, contention, and work." His attention to emotional states and activities would make him a good user experience designer.

So, taking Leonardo as our inspirational muse, we can wonder how his thinking would influence our use of technology. How might Leonardo's integrative approach that blends science and art lead us to new technologies, applications, and designs?

These questions might guide you in thinking about how a truly beneficial technology might reshape your life. They steered me to examine how I could apply a user-centered view to accelerate technology evolution. Such questions could propel developers to get past old ways of thinking that are stuck on the old theme of making impressive computers. They would do better by considering new ways of thinking about facilitating and empowering users. The key questions are not whether broadband wireless networks will be ubiquitous, but how your life will change as a result of them. Life choices come first, technology second. Enduring values should govern technology evolution.

ABOUT THIS BOOK

To reach the goal of promoting human values, we need to build a solid foundation that supports human needs and aspirations. The supporting technology foundation begins with better designs that generate better user experiences with common tools such as word processors, e-mail, and Web pages. Current designs are often too difficult to use. Too many users experience anxiety and frustration when their computers crash, when they can't open e-mail attachments, and when they inadvertently lose their last hour of work. Faster processors and higher bandwidth networks will not save the day—too many designs are unusable at any bandwidth (see chapter 2). So the first step toward the new computing will be to promote good design by getting angry about the quality of user interfaces and the underlying infrastructure. This public outcry can pressure industry leaders and designers to improve their software designs in applications such as word processors and the reliability of support environments provided by operating systems and networks. These changes will accelerate learning and performance while reducing confusing dialog boxes, frustrating crashes, and incompatible data formats.

The second step toward the new computing is inclusiveness, what I call *universal usability,* enabling all citizens to succeed in using information and communication technologies to support their tasks (see chapter 3). This goal leads to designs that support users with new or old computers, fast or slow network connections, and small or large screens. It should make possible participation not only by young and old, novice and expert, able and disabled, but also those yearning for literacy, overcoming insecurities, and coping with varied limitations. Responding to these digital divide challenges will take hard work, but we have come to learn that diversity promotes quality. Accommodating diversity pushes designers to produce higher quality for all users.

If universal usability were achieved, more people could benefit from technology more often. But universal usability is still a dream, a wish, and a hope. The three challenges to designers, managers, and teachers are to support a wide range of technologies, to accommodate diverse users, and to help users bridge the gap between what they know and what they need to know.

Getting adequate public pressure for good design and universal usability during the product development process is the third step. Too often, well-intentioned product managers or software engineers ignore recommendations from human factors and usability professionals. These managers bypass appropriate evaluation processes and choose easier-to-implement design choices. The evaluation processes involve usability testing with real tasks and real users followed by continual monitoring to refine products. The design choices emphasize comprehensible, predictable, and controllable interfaces (see chapter 4).

With these fundamentals in place we can turn to thinking about the future and what we want from the next generation of technology products. Leonardo's inspiration might promote the new computing by encouraging deeper understanding of human activities and relationships. Chapter 5 offers just such a fresh framework for thinking about innovation. It offers an activities and relationships table that could be helpful to users in thinking how to use existing information and communication technology, and to designers who are inventing novel products and services. The columns of the table cover activities such as collecting information, communicating with other people, creating something novel, and sharing it with others. The rows of the table suggest the range of relationships, from intimate friends and family, to colleagues and neighbors, and to broader communities of citizens and markets.

Applying Leonardo's phrase that "work must commence with the conception of man" will push us toward a user-centered design process, with technology on the periphery. Encouraged by Leonardo's interests in learning, his engagement in commerce, his obsession with medicine, and his concern for socially beneficial outcomes, I choose four likely directions for near-term innovations. For each hopeful vision there are many challenges embodied in my provocative rhetorical questions:

> Collaborative education and online courses will be disseminated widely, as universities and companies make face-to-face classroom experiences more intense and expand their audiences with distance and online education (see chapter 6). How can students take greater responsibility for their own education? Why shouldn't teachers have higher expectations for student accomplishment and creativity? Why can't every student earn an A?

> Dramatic shifts in business have already occurred with the emergence of e-business (see chapter 7). Customer relationship management and personalized marketing are indicators of the new opportunities for merchants and customers. Online catalogs, customer service, and purchasing interfaces for e-business, e-services, and e-entertainment will expand rapidly, as will e-complaints. Why shouldn't you get the deal you want?

> The increased responsibility of students is matched by the increased responsibility of patients in the new medicine, sometimes called e-healthcare (see chapter 8). Well-informed patients who seek specialized treatments or participation in clinical trials are challenging the dominant position of physicians and healthcare providers. Two-way telemedicine and healthcare information resources will grow dramatically as patients, nurses, doctors, and health management organizations all go electronic. Why shouldn't advanced networks with adequate privacy protection enable your medical records to be available in every emergency room? Why shouldn't your physician create a special treatment plan for you? Why should you ever be sick?

> The new politics are apparent in more potent public interest groups, livelier political discussion groups, and greater access to government officials. The rapid push toward e-government services will make it easier for people to search vast government digital libraries, influence legislation, and apply for disaster relief funds (see chapter 9). Political deliberation will promote advanced designs that could support rational discourse among millions of citizens while minimizing disruptions. How can citizens make governments even more responsive to their needs while preventing government bloat and unnecessary regulation? How can citizens be heard? Why shouldn't you get what you want from government?

These four applications—e-learning, e-business, e-healthcare, and e-government—take care of basics, but many other applications are also important. We could go further with e-entertainment, e-travel, e-justice, and e-everything, but I hope readers will be able to extrapolate.

An ambitious goal for the new computing is to support your creativity in many domains: sciences and the arts, composing and performing, and work and entertainment. Computers won't ever have Aha! moments; only people are capable of experiencing that joy. However, computers will support your access to previous work, consultation with peers and mentors, rapid generation and exploration of proposed solutions, and dissemination within the field (see chapter 10). They can help make more people more creative more of the time.

The pull of creativity is strong because the satisfactions and rewards can be large. The struggle to solve a problem can be frustrating, but the thrill of success is often proportional to the intensity of the struggle. For some people, the urge to create is so strong that life is unfulfilling if they cannot create. An old Greek aphorism captures this strong connection in a positive way: "Art is life; Life is art."

The University of Chicago psychologist Mihaly Csikszentmihalyi (1996) uses the term *flow* to describe the engaging experience of responding to appropriate challenges: Your absorption is total, the world disappears, time is irrelevant, and your skill is applied entirely to writing a song, making pottery, or playing basketball. It is a thrill!

Creativity support tools can help novices perform at the level of experts, and enable experts to innovate more ambitiously. They expand the possibilities for artists, musicians, poets, playwrights, and journalists to sketch bold ideas, compose fresh symphonies, and write compelling poems. Creativity tools enable scientists, engineers, architects, physicians, and lawyers to analyze more deeply, design more thoroughly, and disseminate more widely. They allow you to do your job as a teacher, student, manager, and salesperson in ways that give you greater freedom to reflect, integrate, and produce. Creativity tools support exploration, discovery, innovation, invention, and more. In the words of *Star Trek,* the goal for many people is "to boldly go where no person has gone before."

Such bold and broad expectations are difficult to satisfy, so the vision I offer in this book will be an enduring challenge. The good news is that existing software provides a good foundation to build on. Of course, there are many problems with contemporary software that need to be overcome, and change is difficult. I try to lay out the possibilities and provide a framework for the next generation of user experiences. The same framework gives you the concepts by which you can organize your work with existing tools. I hope to convince you to join in pushing developers to provide needed improvements.

The book closes with questions and proposals to help us reach still grander goals. Can technology be designed to support peaceful outcomes, conflict resolution, or violence reduction? Can computers support reflection as well as action? self-awareness as well as compassion?

THE SKEPTIC'S CORNER

Advanced technologies have the potential to promote positive contributions, but they also can support the dark side of human nature. Information and communication technologies have been used to disseminate hateful and racist messages. They enable users to spread lies and encourage prejudice. They can alienate children from families, violate privacy, and spread pornography.

Poorly designed information and communication technologies cause frustration, confusion, and anger, as well as contribute to social exchanges marked by hostile comments. Technology flaws have caused deadly errors in medical care, troubling delays in air traffic, or disruptive losses of data and services. Networking has benefits but also allows computer viruses to spread across our vulnerable networks and error messages that warn of a failed hard drive that has entombed all our data.

These unhappy realities do not have to remain forever a part of our technology experience. User groups can pressure manufacturers, developers, and suppliers of information and communication technology to build better environments, just as they have pressured industry to build safer automobiles and more environmentally friendly factories. Skeptics do not believe that the course of technology can be changed. They see competitive market forces and malevolent corporate power to be unstoppable. Dystopian critics fear that things will get worse. They fear that society will fragment along ethnic lines, digital divides will grow across economic gaps, and freedom of choice will recede.

Without a fundamental change in values inspired by a Leonardo-like blend of human-centered design, aesthetics, and engineering, there is a danger that future information and communication technologies could further raise barriers between disciplines and widen the gulf between diverse communities. Without an inspirational muse like Leonardo, poor designs could increase user frustration, erode emotional contacts and undermine empathetic encounters. Without Leonardo's "spiritual kinship for the underprivileged" (Frere 1995, 9) computers could become tools restricted to highly trained specialists and small elite groups. Without Leonardo's clarity of thinking, computers could become complex machines with unpredictable agents launched by confused users. Some users would benefit by being able to master the complexity and overcome the barriers, but most users would have little control and flexibility to explore alternatives or pursue their dreams. As the perceived complexity of technology grows, user empowerment fades. As unpredictability proliferates, user responsibility deteriorates. Users could become victims of the machine.

These dark scenarios could be avoided by giving more attention to the muse of the new computing. The Leonardo muse would be on the side of clarity, simplicity, and beauty. Leonardo's Renaissance spirit, one that combined science and art, could influence the evolution of the new computing in a way that blends advanced technologies with human concerns. The new computing could support creative endeavors while accommodating varied working styles. It could promote participation by culturally diverse users whose complementary knowledge and skills contribute to more creative solutions. I will not promise that every user will become a Leonardo, but for those of you who seek to be more creative and want to build a better world, technology can be a remarkably helpful tool.

Skeptics may also argue that changing the dominant values among technology developers from the old to the new computing is impossible. It is not easy, but the evidence of recent years is that user-centered designs can be the winning strategy.[3] Another challenge is that proposing high-minded concepts about improving quality and labeling them the new computing is the simple part, but the realities of software implementation are much more difficult. This is true. I do not underestimate the challenge nor the good intentions of software professionals, but too often the goals of improved quality are given too little attention.

Finally, even if these tools are wonderful and helpful, low or no technology may be the wiser choice in many cases. The therapeutic benefits of walking in the woods, holding a baby, and talking to your friends should always be respected. Natural surroundings, solitary reflection, and intimate caresses are also important human needs.

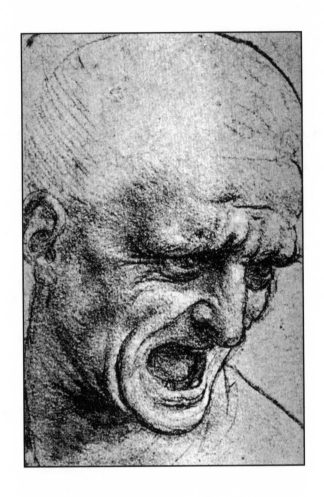

Study of the Head of a Man Shouting, from The Battle of Anghiari. From license-free "Leonardo da Vinci: Selected Works," Planet Art.

2

> UNUSABLE AT ANY BANDWIDTH

A fuchsia cell phone might be pretty.
But a cell phone that does not require a manual—now that is beauty.
—Katrina Galway, Letter to the Editor, *Time Magazine*, July 24,
2000

RAISING PUBLIC AWARENESS

In 1965 the consumer advocate Ralph Nader created widespread apprehension about the automobile industry with his book *Unsafe at Any Speed: The Designed-in Dangers of the American Automobile.* He reported on the design failures of small cars such as the Chevrolet Corvair, whose terrible safety record was concealed by the car manufacturers. He wrote, "I continually encountered profound reluctance, even fear, to speak out publicly by those who knew the details of the manufacturers' neglect, indifference, unjustified secrecy, and suppression of engineering innovation as regards designing safer automobiles." What shocked Nader was the resistance, even among knowledgeable professionals, to speak out about deadly problems.

Similarly, the environmental activist Rachel Carson provoked profound concern about pollution from insecticides and pesticides with her book *Silent Spring.* (1962). Her efforts raised environmental awareness and helped promote ecological concerns worldwide. A review on Amazon's Web site characterizes it as "a portrait of corporate greed and the arrogance and irresponsibility of control agencies and individual specialists." Carson was also troubled by the unwillingness of people to speak up about a serious health threat.

Software with poorly designed user interfaces sometimes has the same deadly or poisonous effects as car design failures or pollutants. A famous example in software circles involves a computer-controlled radiation treatment device for cancer patients called the Therac-25. This marvel of high-tech engineering was designed to cure patients, but it wound up killing some by delivering deadly doses of radiation in what was described as "the worst series of radiation accidents in the thirty-five-year history of medical accelerators" (Leveson and Turner 1993).

The tragic flaw was a result of complex software that included a poor interface. The design failures resulted in a 61-year-old Georgia woman's getting a massive radiation overdose, possibly one hundred times the normal treatment level. She reported a "tremendous force of heat . . . this red-hot sensation" and told the technician, "You burned me." She got little sympathy from the technician, who told her that it was not possible for the machine to give her a burn. There were no immediate signs of a burn, but the treatment area near her neck felt "warm to the touch." She soon noticed a reddening and swelling on her skin, and eventually the burn signs also appeared on her back, penetrating through her body.

Her radiation burns worsened, she lost her breast surgically, and suffered greatly, but no one was convinced that the Therac-25 or its designers were to blame.

Even after two further serious radiation burn stories in Washington State, the Therac-25 was still in use in a Texas cancer center. A male patient was in for his ninth treatment, so when he got a jolt that felt like an electric shock, he jumped off the treatment table. He developed a painful neck and shoulder, nausea and vomiting, paralysis of his left arm and both legs, and died five months later. Again, no one could believe that the Therac-25 had design flaws; especially since the software logs of the equipment usage did not reveal the problem. So after some investigation of the equipment, it was again approved for use. Although training manuals were incomplete, feedback to the operator was poor, and obscure error messages (such as "Malfunction 54") were the norm, everyone accepted that this is what advanced technology was like. Soon after, a second patient in Texas was given a deadly overdose.

The U.S. Food and Drug Administration (FDA) became actively involved, and some fixes were made. However, back in Washington State, another Therac-25 patient soon felt the strange burning sensation of a deadly overdose. Finally, the FDA ordered a shutdown of all Therac-25s and carried out an extensive investigation. A professional journal report on this terrifying case recommended that

> Documentation should not be an afterthought.
> Software quality assurance practices and standards should be established.
> Designs should be kept simple.
> Ways to get information about errors should be designed into the software from the beginning.
> The software should be subjected to extensive testing.

These seem like obvious recommendations, but too often the self-confidence of designers exceeds their concerns for safety. Peter Neumann's remarkable archive of frustrating computer failures, deadly breakdowns, or costly disruptions may encourage more appropriate caution and greater diligence in designing new technologies.[1] Some of the stories are tragic, such as the shooting down of the civilian Iran Air flight 655 over the Persian Gulf in July 1988, which killed 290 people. Poor design of the user interface for the Aegis air defense system contributed to the operator's belief that the Airbus was an attacking fighter plane, and the captain of the *USS Vincennes* consequently ordered a tragically mistaken missile launching.[2] Neumann's archive of thousands of incidents should be required reading for every technologist—it's frightening.

UNUSABLE INTERFACES

As information and communication technologies become a larger part of everyone's life, there are serious dangers that need substantial attention. The first step towards the new computing is to raise awareness and then generate action that yields dramatically improved user experiences with information and computing technologies.

For applications in which life and death are at stake, such as medical, military, transportation, and power systems, there is clearly a requirement for careful design, thorough testing, and continuous monitoring of usage. Even when systems are not life-critical, failures can be costly, wasteful, terrifying, and frustrating. Too often computer software, with chaotic screen layouts, confusing terminology, and incomprehensible instructions, is just too hard to figure out. We've all seen Web pages with bizarre colors that mislead you, distracting animations that confuse you, and blinking advertisements that disrupt your concentration. More unsettling are the intricate navigation paths you have to follow to reach almost hidden features and the infuriating breakdowns in networked communications. Users often feel trapped by an unintelligible dialog box that demands a yes/no choice and long waits for Web pages.

But the users' troubles don't end here. Still greater frustrations await those who get coded error messages like this one:

> This program has performed an illegal operation and will be shut down.
> If the problem persists, contact the program vendor.
>
> KERNEL32 caused a general protection fault
> in module USER.EXE at 0003:000035f6.
> Registers:
> EAX=00000010 CS=1757 EIP=000035f6 EFLGS=00000202
> EBX=013f000b SS=1cf7 ESP=00008f6a EBP=00008f80
> ECX=00020204 DS=165f ESI=81800000 FS=0167

EDX=0001ffed ES=0000 EDI=81700000 GS=0000
Bytes at CS:EIP:
26 08 01 80 4c 0c 01 e9 c1 fb 83 7e f8 00 75 21
Stack dump:
15bf1667 e0d88180 00000001 00010000 000260a4 8faa5c80 01d71346
00bf05cd 80011596 01015c5c 165f03de 176707c2 013f6948 00000000

Ironically, just as I was working on this chapter, the word processor crashed with this message:

WinWord.exe Application Error
The instruction at "0x30a91745" referenced memory at "0x00000407"
The memory could not be "read"
Click on OK to terminate the application
Click on Cancel to debug the application

This left me with little choice but to terminate, reboot, and spend an hour figuring out what had been lost.[3]

While anecdotes abound and user frustration is widespread, there is disturbingly little documentation of how serious the problem is. Air travelers have come to expect regular reports about which airlines or airports have the most delays, and it seems that these reports do influence efforts to improve service. Similarly, reports on hospital surgical records, post office delivery delays, and automobile safety performance are common topics on the evening news. These reports can be quite detailed, getting into which kinds of surgery are best done at which hospital, or how well rear bumpers fare in low, medium, or high-speed crashes.

In a rare survey of six thousand computer users, California-based SBT Corporation found an average of 5.1 hours per week wasted in trying to use their machines. According to this survey, users waste more time in front of computers than on highways. However, we don't know which applications from which suppliers cause the greatest problems, nor what kinds of problems are most prevalent. Surveys are a good start, but observing users and logging usage would provide still more accurate results. Then consumers could choose the better products, and companies would have useful feedback to improve their designs.

Some information from small studies is available. During HomeNet, a well-financed but controversial study of forty-eight Pittsburgh area homes, 133 participants received computers, free network connections, training, and assistance with problems (Kraut et al. 1996). Even in such optimal conditions a central limitation was the difficulty that users experienced with the services. The researchers wrote, "Even the easiest-to-use computers and applications pose significant barriers to the use of online services . . . even with help and our simplified procedure, HomeNet participants had trouble connecting to the Internet."

The University of Colorado psychologist Tom Landauer (1995) describes "the trouble with computers" in his well-documented arguments for user-centered design. He highlights the economics of poor user interface design with case studies of how effort invested in design paid off in reduced life cycle costs. Similar reports from IBM claimed 100 to 1 returns on investments in usability testing because of the reduced training, maintenance, production, and revision costs. (Karat 1994). The business case for usability has been made repeatedly and effectively.

The analogy between auto safety and computer usability was noted by two MIT researchers, Christopher Fry and Henry Lieberman (1995): "Most current programming environments are like the old Corvair; when an error occurs, the programmer is left staring at nothing but the flaming wreckage of an error message and perhaps a core dump. Every program crash totals the vehicle, and nothing can be learned from the experience except to buy a new car and start over."

Their harsh characterization may reflect the frustrations of many users, programmers, and consumers of the old computing. However, new computing methods can produce more usable, more reliable computer software and user interfaces that yield much improved user experiences.

GETTING TO THE NEW COMPUTING

So how might we take steps to promote the new computing? There is no magic bullet that will bring widespread use of low-cost devices that are easy to learn, rapid in performing common tasks, and low in error rates. The main change that is needed is not a technology breakthrough. The most important breakthrough will be your change in expectations and willingness to ask for higher quality. Consumer pressure would push companies to produce improved designs that are more reliable, learnable, and usable. Then improved testing would eliminate more of the problems, and quality control would become an obsession. When the software was shipped, it would be more likely to work reliably.

Emphasizing high quality is probably just what Leonardo would do. He concentrated on the details for his *Last Supper,* making sure all the lighting angles fit, the garments hung realistically, and each face expressed the appropriate emotion. Leonardo's pursuit of quality is also visible in his portraits *Mona Lisa* and *Ginevra de'Benci.* He refined painterly techniques for blurring backgrounds to give a remarkable sense of depth and knew where to pay attention to details such as drawing each strand of hair separately. He also understood mathematical perfection, as his *Vitruvian Man* (figure 2.1) exemplifies; the drawing incorporates several mathematical proportions enunciated by the architect Vitruvius.[4] Painterly precision and mathematical perfection would be good traits for modern software engineers and interface designers. Do we yet have modern Leonardos of software, admired for their beautiful designs and reliable programs? How might we encourage, recognize, and reward such geniuses? Whose training studios are likely to produce software Leonardos?

Of course there are differences between a painting and a software product. The modern goal of building beautiful software is a substantial challenge requiring close coordination among hundreds of people. It is not an easy job, and the devoted groups of designers and engineers who work for leading software companies deserve respect. But consumers deserve respect, too. The current level of problems should not be acceptable to consumers, and their pressure to get what they need could encourage industry managers to provide further resources so that higher quality can be assured.

A computer users' movement could also influence independent consumer organizations and government agencies to more thoroughly evaluate products and measure user experiences. Automobile manufacturers must report problems and offer repairs and recalls to their customers if problems are found, even years after delivery. Independent testing organizations, insurance companies, and government agencies conduct automobile crash tests to verify safety requirements, such as rollover resistance. Software suppliers could be pressured by a public outcry to

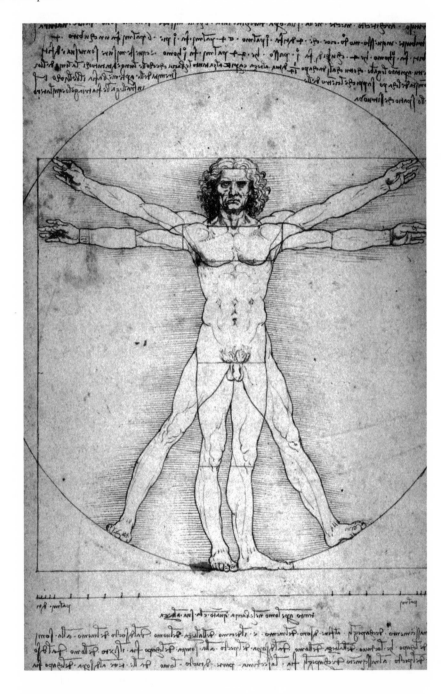

2.1   Leonardo's drawing *Vitruvian Man* (Man of Perfect Proportions). From license-free "Leonardo da Vinci: Selected Works," Planet Art.

report on usage problems, make fixes, or pay damages, just as automobile manufacturers do.

One idea would be to encourage software suppliers to provide a one-dollar reward for reporting your software crashes or a dime for dialog boxes you don't understand. This could be arranged by e-mail and credit given for future software purchases (figure 2.2).

Of course there are problems with this proposal, but it is meant as a provocation to new ways of thinking. Users would not get credit for reporting the same problem more than once, and patterns of users getting repeated credits would have to be stopped, but the basic idea is to provide thorough reports about customer problems. It also reflects a sense of fair play when things go wrong. When a waiter spilled tomato soup on my pant leg, he apologized, offered to pay for dry cleaning, and gave our table free desserts. When an airline delay for mechanical problems exceeds a certain time period, passengers are entitled to free meals or hotels. Software suppliers should be expected to take responsibility for their users, too.

When software suppliers hear this proposal, they often respond negatively and argue that it is unworkable or too expensive. It is encouraging that Netscape has added a Quality Feedback System that enables users to send an e-mail report (though receiving no compensation) when its software crashes.[5] Software suppliers who worry about the cost should realize the benefits of improved quality. What is really too expensive is the cumulative cost of the wasted efforts of tens of millions of users when software crashes. If each user spends only ten minutes per day recovering from a crash or finding a work-around for other failures, that makes about an hour per week, or about a week per year wasted. This is far lower than the 5.1 hours reported in the SBT survey, but if we estimate a week's salary (including overhead) for computer users at $500 each, multiplied by a 100 million users in North America, this alone yields $50 billion per year. Even if the real cost were half this amount, the drag on the economy is large, and it increases if we count worldwide users.

Software suppliers sometimes argue that software is complex and inherently buggy, so the supplier can't be expected to be perfect. This argument may have been acceptable twenty years ago, but as software matures, shouldn't we, as users, expect that the core functions of common packages are guaranteed and reliable. Unfortunately, many software suppliers are moving in the opposite direction, seeking protection from

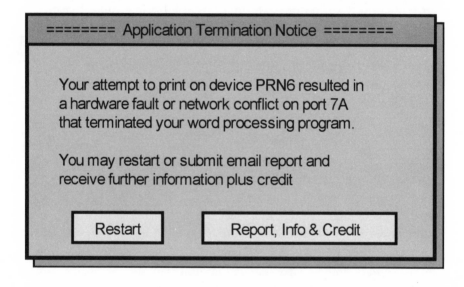

2.2   Proposed dialog box that offers users a chance to report product failure and get credit for future purchases.

consumer complaints and writing software license agreements that strip away even basic consumer protections. Consumers should be aware of the need to promote their rights against proposals that would allow software suppliers to avoid liability for damages, change license terms at any time, prohibit negative reviews, and access your files so that they could remove software if they determine you have violated the license.

Promoting a consumer uprising, as there was with automobiles and the environment, seems difficult but necessary. It is time to get angry about the quality of user interfaces. It is time for you to complain to software suppliers when products fail, lobby government representatives to provide consumer protection, and push industry leaders to provide stronger warranties. Such efforts have yielded constructive results in automobile safety, tobacco control, and environmental concerns. How might we get prominent legislators, respected journalists, or consumer activists to step forward?

Another useful step would be to reward and honor those who have done a good job. Quality awards have appeared in many fields, and should be promoted for information and communication technologies. A good start in this direction is the Webby award for excellence in Web sites.[6] This award covers content, structure and navigation, visual design, functionality, interactivity, and overall experience. The annual Webby awards ceremony brings some glamor and thoughtful analysis to Web design. But how about awards for ease of learning, clarity of instructions, online help, error prevention, and error messages? And why not awards for best customer support service?

Of course, many other steps are needed to make the new computing a golden age of usability. Industry groups need encouragement to voluntarily develop strategies to ensure, measure, and warrant usability and high quality. Independent consumer agencies could contribute by more regularly evaluating and endorsing products. Government agencies such as the Federal Trade Commission and the Federal Communications Commission have been effective in studies of Web site privacy policies, child pornography violations, and the persistence of the digital divide.

A fundamental change in education would be to train teachers so that they knew enough about computing to use and teach it competently. Such a program would go well with a plan that established target skill levels for computer use, especially for middle and high school students.

Expanded government support for university research and education in human-computer interaction would make possible accurate data collection about the severity of the problems, train a new generation of users, and support pilot projects that demonstrate improvements. Each of these changes would help shift public attitudes and encourage industry to give greater attention to users and their problems.

THE SKEPTIC'S CORNER

Skeptics say that you can't change the way industry leaders and technology developers behave. This negative attitude can be self-fulfilling and is used by some people to avoid change. Positive steps are needed because there are strong forces that resist the shift from the old computing to the new computing, for instance, those on the following list (adapted with permission from Mehlenbacher 1999):

> Engineers who refer to usability issues as "soft" or to users as "stupid"
> System designers who believe that soon we'll have an automatic interface design program
> Usability "specialists" who superficially edit product releases as they're about to be shipped
> Designers who've never met or talked to someone who actually uses their products
> Managers who maintain that all their software developers already care about usability issues
> Marketers who argue that most users don't know what's good for them in the first place
> Consumers who continue to do business with companies that produce unusable products

However, Ralph Nader showed that the car industry, government agencies, and the public can become safety-conscious. Similarly, Rachel Carson elevated environmental consciousness to a highly political level so that every candidate must have an environmental policy that responds to public concerns about pollution, destruction of forests, or toxic wastes.

Cynics also argue that usability is a good notion, but that perfection of the underlying technology is necessary before applying user-centered thinking. Putting technology first is definitely the old computing, but market successes in the new computing may change the way designers work. Another cynic noted that the road to technology-centered solutions is paved with user-centered intentions. The experience of many usability professionals has been that they were invited to participate in projects, but then their advice was ignored. When the products failed, the usability professionals were doubly frustrated because they got blamed.

These darker scenarios are part of the early stages of the transformation to the new computing. As time goes by, usability practices and human-computer interaction (HCI) research are spreading. Usability heroes and HCI gurus are becoming prominent and more numerous. Corporations like America Online make the point that their interface is "So easy to use, no wonder it's #1." They recognize the centrality of usability and have done much to make their services usable to a wide range of people.

An Older Man and a Younger Facing One Another. From
license-free "Leonardo da Vinci: Selected Works," Planet Art.

3

> ## THE QUEST FOR UNIVERSAL USABILITY

I feel . . . an ardent desire to see knowledge so disseminated through the mass of mankind that it may . . . reach even the extremes of society: beggars and kings.
—Thomas Jefferson, Reply to American Philosophical Society, 1808

DEFINING UNIVERSAL USABILITY

An important step toward the new computing is to promote the compelling goal of universal access to information and communication services. Enthusiastic networking innovators, business leaders, and government policymakers see opportunities and benefits from widespread usage. But even if they succeed in bringing low costs through economies of scale, new computing professionals will still have much work to do. They will have to deal with the difficult question, How can information and communication technologies be made usable for every user?

Designing for experienced frequent users is difficult enough, but designing for a broad audience of unskilled users is a far greater challenge. Scaling up from a listserv for 100 software engineers to 100,000 schoolteachers to 100,000,000 registered voters will take inspiration and perspiration.

Users of information and communication technologies also have a vital role to play in pushing for what they want and need. Customer-oriented pressure will accelerate efforts to make new computing technologies usable and useful. Older technologies such as postal services, telephones, and television are universally usable, but computing technology is still too hard for too many people to use. Low-cost hardware, software, and networking will benefit many users, but interface and information design improvements are necessary to achieve higher levels of success.

We can define universal usability as more than 90 percent of all households being successful users of information and communication technologies at least once a week. A 2000 survey of U.S. households shows that 51 percent have computers and 42 percent use Internet-based e-mail or other services (NTIA 2000), but the percentages decline in poorer and less educated areas. Mario Morino's leadership in promoting technology access and social development in low-income communities provides a realistic vision based on ten premises. His 2001 report, "From Access to Outcomes: Raising the Aspirations for Technology Initiatives in Low-Income Communities," encourages working through trusted community leaders to build community capacity with affordable housing, health clinics, public transportation, and other services. He recognizes the catalytic power of e-mail lists, the importance of skills training, and the value of improved software design.

Internationally, meeting the challenge of universal usability is still harder. Internet dissemination is much lower in other countries around the world than in the United States. Many European countries have Internet usage rates approaching 50%, but in South America, the leading Internet-using country is Brazil, with only 3% usage. In many African and Asian countries there is only one Internet Service Provider, and usage is well below 1% of the population.[1] The United Nations Development Programme and the United Nations Information Technology Service seek to apply information technology for community building at an international level.[2] They coordinate activities with groups such as British Partnerships Online, which is devoted to developing information community centers, and Volunteers in Technical Assistance (VITA), which promotes appropriate communication, agricultural, and manufacturing technologies in Benin, Mali, Guinea, and other developing countries. Cost is a central issue for many, but hardware limitations, perceived difficulty, and lack of utility discourage others. It is hard to overstate the importance of addressing international digital divides because of the potential for accelerating economic development that benefits all nations and the opportunity to promote social initiatives that support constructive, rather than violent, movements. If countries are to meet the goal of universal usability, then researchers and technology developers need to aggressively improve current products, tune them to the realities of local needs, and increase the relevance of Web services.

Leonardo would probably be a promoter of universal usability. He was described as having "a spiritual kinship for the underprivileged" (Frere 1995, 59) and sought to please and serve the needs of rich and poor. Many of his mechanical inventions, public artworks, and urban plans were meant to benefit the full spectrum of Florentine and Milanese citizens. His weapons and defensive fortifications supported urban protection from invaders. His theatrical constructions and clever toys are other examples of Leonardo's desire to please a wide range of people. Leonardo was not a scholar isolated in a university or a scientist dealing with esoteric theories. He may have worked for the nobility, but he was very much a man of the people, wandering through the Florentine markets and sketching ordinary citizens as well as painting portraits for the nobility.

Our modern understanding of universal access is usually linked to the U.S. Communications Act of 1934 covering telephone, telegraph, and radio services. It sought to ensure "adequate facilities at reasonable charges," especially in rural areas, and prevent "discrimination on the basis of race, color, religion, national origin, or sex." The term *universal access* has been applied to computing services, but the greater complexity of computing services means that access is not sufficient to ensure successful usage.

Therefore *universal usability* has emerged as an important issue. The complexity of information and communication technologies stems, in part, from the high degree of interactivity that is necessary for information exploration, commercial applications, and creative activities. The Internet is compelling because of its support for interpersonal communications and decentralized initiatives: entrepreneurs can open businesses, journalists can start publications, and citizens can organize political movements.

The increased pressure for universal usability is a happy by-product of the growth of the Internet. Since communication, e-business services (shopping, finance, and travel), e-learning, and e-healthcare are expanding and users are becoming dependent on them, there is a strong push to ensure that the widest possible audience can participate. A particularly strong argument for universal usability is tied to e-government applications such as access to national digital libraries and the movement towards citizen services at federal, state, and local levels. These services include tax rules and filing, Social Security benefits, foreign travel information, commercial licensing, recreation and parks, and police and fire departments. Another circle of support for universal usability includes daily life needs provided by employment agencies, training centers, parent-teacher associations, public interest groups, community services, and charitable organizations.

Critics worry about the creation of an information-poor minority, or worse, Internet apartheid. Although the digital divide (Campaine 2001) in Internet usage has been shrinking between men and women, and between old and young, the digital divide is growing between rich and poor (NTIA 2001). (See figure 3.1.) Well-off households are three times as likely to have Internet access as poor households. Similarly, education levels produce a digital divide between well and poorly educated households. (See figure 3.2.) Less well documented is the continuing separation between cultural and racial groups, and the low rates of usage by disadvantaged users whose unemployment, homelessness, poor health, or

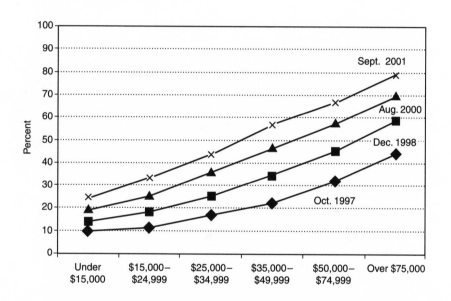

3.1   Percent of U.S. households with Internet access, by income ($000s), 1998 and
2000. *Source:* NTIA 2001.

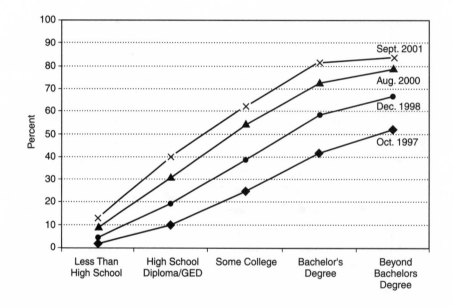

3.2   Percent of U.S. households with Internet access, by education, 1998 and 2000.
Source: NTIA 2001.

cognitive limitations raise further barriers. Even more challenging is the situation of international users or refugees whose literacy skills may be poor and whose languages may not be well represented on the World Wide Web. Improved design and multilingual capabilities can do much to reduce the digital divide, but training at community centers and schools will be important contributors to reducing this divide.[3]

There are other criticisms of information and communication systems. These include concerns about breakdown of community social systems, alienation of individuals that leads to crime and violence, and expansion of bureaucracies. Further threats come from loss of privacy, inadequate attention to communications or power breakdowns, and exposure to malicious viruses or hostile attacks. Open public discussion of these issues by way of participatory design strategies and town meeting forums can help to improve proposals and win public support. I have proposed that significant plans for new technology applications by government agencies should be accompanied by publication of social impact statements (Shneiderman and Rose 1996). These documents, modeled after environmental impact statements, are written to make them accessible to the public and might elicit widespread awareness and diverse proposals that reduce negative and unanticipated side effects.

Technology enthusiasts can be proud of what they have accomplished and of the number of successful Internet users, but deeper insights will come from understanding the problems of frustrated users and of those who have stayed away. Each step to broaden participation and reach these forgotten users by providing useful and usable services will be welcomed.

Universal usability is sometimes tied to meeting the needs of users who are disabled or work in disabling conditions. This important direction is likely to benefit all users. The adaptability needed for users with diverse physical, visual, auditory, or cognitive disabilities is likely to benefit users with differing preferences, tasks, skills, hardware, and so on. The recent growth of interest in disability access is tied to the standards for "comparable access" by disabled users, as specified under section 508 of the U.S. Rehabilitation Act as amended by Congress in 1998. (Access Board 2000). The beneficial effects of changes to Web sites and other technologies may spill over to bring positive changes for all users.

Advocates who promote accommodation of disabled users often describe the curbcut—a scooped-out piece of sidewalk to allow wheelchair users to cross streets. Adding curbcuts after the curbs have been built is expensive, but building them in advance reduces costs because less material is needed. The benefits extend to baby carriage pushers, delivery service workers, bicyclists, and travelers with roller bags. Computer-related accommodations that benefit many users are power switches in the front of computers, adjustable keyboards, and user control over audio volume, screen brightness, and monitor position.

Automobile designers have long understood the benefits of accommodating a wide range of users. They feature adjustable seats, steering wheels, mirrors, and lighting levels as standard equipment and offer optional equipment for those who need additional flexibility.

Reaching a broad audience is more than a democratic ideal; it makes good business sense. The case for *network externalities,* the concept that all users benefit from expanded participation, has been made repeatedly. Facilitating access and improving usability expands markets and increases participation of diverse users whose contributions may be valuable to many. Broadening participation is not only an issue of reducing costs for new equipment. As the number of users grows, the capacity to rapidly replace a majority of equipment declines, so strategies that accommodate a wide range of equipment will become even more in demand.

COPING WITH TECHNOLOGY VARIETY

Supporting a broad range of hardware, software, and networks is not an easy task. The job is even more challenging when one considers the need to accommodate old hardware and software as well as new features and environments.

My friend's 93-year-old grandmother is a successful computer user, but she struggles in isolation because of her inability to keep up. She has a 1985 computer with a ten-megabyte hard disk, a character-based green screen, and an ancient word processor. She types and prints letters, but getting connected to e-mail would require a lot of changes. Adequate help is hard to get, and no company supports her technologies. Her grandson comes by every so often and gives her a hand, which does create a nice bond between them, but moving up to newer technology seems too difficult.

Every user of computers has to decide about keeping up with change. The new features can be attractive, but the fear of upgrades has become a national source of anxiety. Most users have stories of how their last upgrade caused unexpected failures or how it took them weeks to convert files. The stabilizing forces of standard hardware, operating systems, network protocols, file formats, and user interfaces are undermined by the rapid pace of technological change. Technology developers delight in novelty and improved features. They see competitive advantage in advanced designs, but these changes disrupt efforts to broaden audiences and markets. Limiting progress is one solution, but a more appealing strategy is to pressure developers to make information content, Web services, and user interfaces more malleable and adaptable to change.

The range of processor speeds in use varies by a factor of 1,000 or more. Moore's law, which states that processor speeds double every eighteen months, means that after ten years the newest processors are one hundred times faster than the old ones. Designers who wish to take advantage of new technologies risk excluding users with older machines. Similar changes for random access memory (RAM) and hard disk space also inhibit current designers who wish to reach a wide audience. Other hardware improvements such as increased screen size and improved input devices also threaten to limit access. Accommodating varying processor speed, RAM, hard disk, screen size, and input devices could help cope with this challenge. Shouldn't software be designed so that users could run the same calendar program on a palm-sized device, a laptop, and a wall-sized display?

Software changes are also a concern. As applications programs mature and operating systems evolve, software becomes obsolete because newer versions fail to preserve file format compatibility. Some changes are necessary to support new features, but modular designs are needed that promote evolution while ensuring compatibility and bidirectional file conversion. The Java movement is a step in the right direction, since it proposes to support platform independence, but its struggles indicate the difficulty of the problems.

Another concern is the variety of network connection speeds. Some users will continue to be limited to slower telephone dial-up modems while others will use much faster cable or DSL modems. This hundredfold differential creates an enormous chasm between user communities. Since many Web pages contain large graphics, user control of byte counts would be a huge benefit. Most browsers allow users to inhibit graphics, but more flexible strategies are needed. You should be able to select information-bearing graphics only or reduced byte count graphics, and invoke procedures on the server to compress the image from 300K to 80K or to 20K.

Another development that is needed is software to convert interfaces and information across media or devices. If you want your Web page contents read to you over the telephone, as many blind users do, there are already some services.[4] However, improvements are needed to speed delivery and extract the content appropriately. A more advanced idea, a generalization of the Universal Serial Bus, is a total access system that would allow you to attach a wider array of input or output devices to a computer.[5] This would enable users with disabilities or special needs to connect their specialized equipment to any computer, just as they bring along their own eyeglasses or hearing aids.

ACCOMMODATING DIVERSE USERS

Users vary with respect to computer skills, knowledge, age, gender, disabilities, disabling conditions (mobility, sunlight, noise), literacy, culture, and income.

Since skill levels with computing vary greatly, search engines usually provide a basic and an advanced dialog box for query formulation. As a novice you can proceed without too many impediments, whereas experts can fine-tune their search strategy. Since knowledge levels in an application domain vary greatly, some sites provide two or more versions of the content. For example, the National Cancer Institute provides introductory cancer information for patients and in-depth details for physicians.[6] Since children differ from adults in their needs, NASA provides a children's section on its space mission pages.[7] Universities often segment their sites for applicants, current students, or alumni but then provide links to shared content of mutual interest.

Similar segmenting strategies could be applied to accommodate users with poor reading skills or users who require other natural languages. While there are some services to automatically convert Web pages to multiple languages,[8] the quality of human translations is still higher. If an e-commerce site maintains multiple-language versions of a product catalog, then it should make simultaneous changes to product price (possibly in different currencies), name (possibly in different character sets), or description (possibly tuned to regional variations).

As a consumer you should expect Web sites to accommodate your needs depending on your interests, income, cultural background, or religion. You should be able to find music, food, or clothing catalogs attuned to your needs so that you can easily find the products you want and not be offended by things you are not interested in. You'll probably find yourself revisiting the e-commerce sites that follow these strategies. If you are looking for Mozart symphonies, you shouldn't have to page through long lists of B. B. King songs and vice versa.

For disabled users, the needs are even more critical—if the Web site designer has not accommodated their needs, they are unhappy visitors or lost customers. Many systems allow partially sighted users, especially elderly users, to increase the font size or contrast in documents. This is good news, but a complete solution includes allowing users to improve readability in control panels, help messages, or dialog boxes. Blind users will be more active users of information and communication services if they can receive documents by speech generation or in Braille, and provide input by voice or through their customized interface devices. Physically disabled users will eagerly use services if they can connect their customized interfaces to standard graphical user interfaces, even though they may work at a much slower pace. Cognitively impaired users with mild learning disabilities, dyslexia, poor memory, and other special needs could also be accommodated with modest changes to improve layouts, control vocabulary, and limit short-term memory demands.

Expert and frequent users also have special needs. Enabling customization that speeds high-volume users, shortcuts to support repeated operations, and special-purpose devices could improve interfaces for all users. Such expert or professional customizations may represent an important business opportunity.

Finally, appropriate services for a broader range of users need to be developed, tested, and refined. Corporate knowledge workers are the primary audience for many contemporary software projects, so the interface and information needs of the unemployed, homemakers, the disabled, or migrant workers usually get less attention. This has been an appropriate business decision till now, but as the market broadens and key societal services are provided electronically, the forgotten users must be accommodated. For example, Microsoft Word provides templates for marketing plans and corporate reports, but "every-citizen" interfaces might help with job applications, babysitting cooperatives, or templates for letters to City Hall. And what about first-aid, 911 emergency assistance, crime reporting, or poison control on the Web? We should expect these services. In disaster and crisis situations, Internet services may be more reliable than telephones, but little attention has been devoted to such needs.

The growth of online support communities, medical first-aid guides, neighborhood-improvement councils, and parent-teacher associations will be accelerated as improved interface and information designs appear. Community-oriented plans for preventing drug or alcohol abuse, domestic violence, or crime could also benefit from improved interface and information designs. Such improvements are especially important for government Web sites, since they are moving toward providing basic services such as driver registration, business licenses, municipal services, tax filing, and eventually voting. Respect for the differing needs of users will do much to attract them to advanced technologies.[9] As citizens we should demand good service for all.

BRIDGING THE GAP BETWEEN WHAT USERS KNOW AND WHAT THEY NEED TO KNOW

Every user of computers must learn how to use interfaces to accomplish his or her task. Whether you are trying to manage your retirement funds or find an apartment in a new city, there are new concepts to learn and new information to acquire. What is a margin account? How can I set a stop limit order? How can I get a map to show me how far it is from Lincoln Park to Hyde Park?

The challenge for technology developers is to bridge the gap between what users know and what they need to know. Many users don't know how to begin, what to choose in dialog boxes, how to handle crashes, or what to do about viruses. You should look for software that applies fadeable scaffolding (instructional aids that can be removed as your skills improve), training wheels (limited features to prevent errors for beginners), and just-in-time training (instructions accessible whenever the user runs into trouble).

There are many competing theories about how to train users, but too little study of what really works. One popular theory is the minimal manual approach that suggests reducing up-front instruction and getting users to be active quickly, even if they make some mistakes. This theory is applied in the short "Getting Started" guides for new hardware or software. Two other theories are the constructivist (gets the user to carry out practical projects soon) and social construction (engages pairs or larger groups of users in working together to learn).

Users approach new software tools with diverse skills and multiple intelligences. Some users need only a few minutes of orientation to understand the novelties and begin to use new tools successfully. Others need more time to acquire knowledge about the objects and actions in the applications domain and the user interface. Current interfaces could be improved with more lucid instructions, better error prevention, regular use of graphical overviews, and more effective tutorials for novices. Intermittent users would benefit from more well-designed online help, and experts need compact presentations of guidance materials and shortcuts for frequent tasks. Other helpful aids include easily reversible actions and detailed history keeping for review and consultation with peers and mentors.

A fundamental interface improvement would be support for evolutionary learning and a level-structured approach to design (Baecker et al. 2000). Why can't you begin with an interface that contains only basic features (say 5 percent of the full system) and become expert at this level within a few minutes? Game designers have created clever introductions that gracefully present new features as users acquire skill at the first level of complexity. Could similar techniques apply to the numerous features in modern word processors, e-mail handlers, and Web browsers? A good beginning has been made in some advanced systems with features such as training wheels and "Getting Started" guides, but scaling up and broader application are happening slowly. (Carroll and Carrithers 1984). A good level-structured design in the interface must be accompanied by levels in the tutorials, online help, and the error messages.

Finally, many users might be assisted through online help by way of e-mail, telephone, video conferencing, and shared screens. There is no single best way—you should be able to get help in the way that you find most comfortable at the moment. Some users like to read stories of how other users have solved their problems or used new technologies. For these users, Web sites with case studies, best practices, common problems, and frequently asked question (FAQ) lists are helpful. Many users like to talk about their problems and ask for help in highly social ways involving peers rather than experts. If you are one of these users, you might try chat rooms, news groups, or online communities.

THE SKEPTIC'S CORNER

This chapter focuses on three universal usability challenges: technology variety, user diversity, and gaps in user knowledge. Skeptics caution that accommodating lower end technology and lower ability users, or users with fewer skills will result in a lowest-common-denominator system that will be less useful to most users. This scenario, called "dumbing down," is a reasonable fear, but my experience supports a brighter outcome. I believe that accommodating a broader spectrum of usage situations forces technology developers to consider a wider range of designs and often leads to innovations that benefit all users. For example, Web browsers, unlike word processors, reformat text to match the width of the window. This accommodates users with small displays (narrower than 640 pixels), and provides a nice benefit for users with larger displays (wider than 1,024 pixels), who can view more of a Web page with less scrolling. Accommodating narrower (less than 400 pixels) or wider (more than 1,200 pixels) displays presents just the kind of challenge that may push designers to develop new ideas. For example, they could consider reducing font and image sizes for small displays, moving to a multicolumn format for large displays, exploring paging strategies (instead of scrolling), and developing overviews.

A second skeptics' caution, called innovation restriction, is that attempts to accommodate the low end (technology, ability, or skill) will constrain innovations for the high end. This is again a reasonable caution, but if designers are aware of this concern, the dangers are easily avoidable. A basic Web page accommodates low-end users, and sophisticated user interfaces using Java applets or Flash plug-ins can be added for users with advanced hardware and software and fast network connections. New technologies can often be provided as an add-on or plug-in rather than as a replacement. As new technologies become perfected and widely accepted, they become the new standard. Layered approaches have been successful in the past, and they are compelling for accommodating a wide range of users. They are easy to implement when planned in advance but often difficult to retrofit.

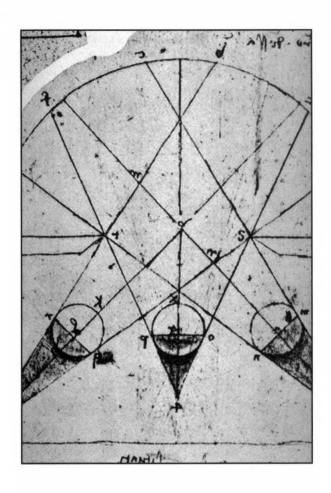

Diagram Illustrating the Theory of Light and Shade. From license-free "Leonardo da Vinci: Selected Works," Planet Art.

4

> NEW METHODS, NEW GOALS

We must not assume that the future of the Internet will be determined by some mindless, external "technological imperative." The most important question is not what the Internet will do to us, but what we will do with it.
—Robert Putnam, *Bowling Alone* (2000), 180

SHIFTING TO THE NEW COMPUTING

Excellence in design (chapter 2) and universal usability (chapter 3) are the first two steps toward the new computing. The third step is ensuring adequate consideration for good design and universal usability during product development. New design methods combined with a fresh set of guidelines and goals could accelerate movement to the new computing.

To add force to this movement, consumers, aided by journalists, politicians, and consumer activist groups, will have to create sufficient pressure on industry leaders to adopt more effective development methods and new goals. In the early days of computing, technology developers built text editors for their own use based on ad hoc methods. Since there was little design knowledge available, they often produced complex commands that required substantial learning and frequent use to achieve mastery. The command to delete a single character might require ten or more characters to type, for example,

```
change /bead/bed/
```

and its successful execution required that an invisible cursor be sitting at the appropriate line of text.

The technology enthusiasts and early adopters were happy enough with these old computing products; they even enjoyed the challenge of learning the multiple options and defaults of Unix commands like "rmdir" and "grep." My favorite example of an obscure command is

```
grep -v ^$
```

which deletes blank lines. Technophiles were proud of having mastered these esoteric environments, achieving guru status as they occasionally helped the poor newbies (newcomers).

The developers of the old computing technology also designed networks to support computer-to-computer exchange of large data files rather than personal messaging. Even the desktop computer hardware was fashioned with gray colors and sharp corners to please the tech-savvy professionals. Technology developers thought about their own needs and used those as the basis for building products that worked adequately for those needs.

Then the user community began to expand to include scientists operating telescopes, financial analysts running economic models, and business professionals writing spreadsheets. Personal computers emerged in 1981, and their popularity grew as graphical user interfaces opened computing to more people.

The appearance of the World Wide Web in the early 1990s changed everything. Easier interfaces and the attraction of information from many sources produced a huge influx of users. The user community quickly broadened to include rebellious artists making provocative animations, on-the-road truck drivers trying to find their destinations, retired nurses investing in mutual funds, and migrant farm workers sending e-mail to their parents.

This amazingly diverse set of users challenge the next generation of technology developers, who are beginning to realize that they must understand what users want. The most astute developers recognize that the audience for computing now includes the late-adopting majority and even technology resisters who must learn about computers to get a job or to download music files. The hard part for technology developers is that they must accommodate these users' low enthusiasm for technology and still lower tolerance for frustration. After all, these users are not like the old computing users who were interested in computers; these new computing users just want to use the technology to get their work done or have some fun.

METHODS FOR ACHIEVING USER-CENTERED DESIGNS

The starting point for a new computing project is to understand who the users are and what they are doing. It's simple to say, but tough to do (Nielsen 1993; Shneiderman 1998). The methods for enabling technology developers to understand users have been expanding in the past two decades.

The first method to support user-centered design is a *user needs assessment* that determines the range of services needed by users. In our work with the Library of Congress (Marchionini et al. 1993), we watched hundreds of users of the online public access card catalog and logged their activities. This well-designed but definitely old-computing style of interface required a one- to three-hour training course. The user needs assessment gave us a clear picture of what were the most frequent operations (such as a search by author name). It revealed how rarely complex Boolean combinations (AND, OR, NOT) were used and how frequently users wanted to see all the books by a given author. We also tracked what questions were asked of the library staff and found to our amazement that the most frequent question was, Where is the bathroom? The librarians quickly put up a sign to answer such questions.

Our user needs assessment gave us important evidence about what to put into a simplified touch-screen interface, such as brief examples of usage ("Hemingway, Ernest") rather than lengthy instructions, and compact displays of search results. The fifty-five terminals, surrounded by elegant wood paneling from the discarded card catalogs, dramatically changed the behavior of visitors who could now accomplish most simple searches on their own without training. It also changed the work life for the Library of Congress staff, who now had time to assist advanced users with more sophisticated queries. To accommodate vision-impaired users, a large screen with special software that magnified the text was installed.

For novel applications or expansions of existing applications, user interviews help to determine what users are trying to accomplish. Ethnographic methods and anthropological theories are in fashion, and they have been quickly adapted to meet the rapid pace of Internet development projects. Ethnographic observation includes videotaping, logging, and simply watching what users do. More active approaches such as participant observation allow interaction with those being studied, but in a careful way that minimizes disruption of normal behavior.

Anthropological theories guide observation and interpretation in ways that lead to more complete and unbiased reports about common practices, beliefs, and relationships. Traditional anthropology is respectful of the cultures being studied and stresses careful efforts to minimize change in those cultures. Computing-related anthropology is closer to applied anthropology because the goal is almost always to change the culture, technology, or work practices. Ethnographic methods have led to important insights about how air traffic controllers collaborate (Hughes 1995) as well as richer descriptions of how teenage girls use computers (Laurel 2001).

Observations for just a few hours can often reveal typical patterns in behavior, common work-arounds to deal with problems, and opportunities to improve collaboration. Interviews, either structured or open-ended, can probe deeper to understand frustrations or satisfactions. For more objective data, software tools that keep a log of user actions can confirm hypotheses about the rhythm of work or play over days or weeks. Sometimes interviewees claim very different patterns of usage than the logs reveal.

When the designers have understood the sequences and frequencies of tasks, they can make multiple mockups of what the interface might be like. These can be hasty sketches, careful drawings, or computer-supported layouts. Increasingly good tools for rapid prototyping are making it easy to create new interface prototypes in a few hours. Such high fidelity prototypes can be interactive and operate almost like the final product. This gives designers, managers, and customers a chance to comment before further effort is invested in making a complete system. Modern programming tools, such as user interface development environments, have made it possible to rapidly build and revise prototypes.

The second method of user-centered design, *usability testing,* is the key to rapid design evolution. The method is simple and has low cost; technology developers ask typical users to carry out realistic tasks with the working prototype and observe where users get into trouble. The laboratory-like setting may compromise reality, but the controlled environment helps focus attention on problems that users encounter with the prototype. The goal is to identify where users get stuck and suggest improvements.

In the most common type of usability testing, three to eight users are asked to work on the task list for one to three hours while they think aloud about their decision-making rationale. The usability testers observe what happens and possibly videotape the events for later review. With most initial designs, the difficulty and resulting anxiety can be quite dramatic as test participants struggle to figure out how to operate a system—an obvious indication of a bad design. The anxiety of usability test participants is lowered when they are told that they are not being tested—it is the prototypes that pass or fail.

The heart of the usability test report is the tabulation of problems users encountered. This is usually followed by recommendations about improvements, possibly organized by low-medium-high priority and low-medium-high effort. Improvements might be to use more consistent and familiar terminology, more constructive error messages, and clearer layouts of information.

Usability testing speeds development and improves quality enormously; Clare-Marie Karat (1994) reports that money spent on usability testing produces a 100 to 1 payoff in reduced costs over a system's lifetime. Some managers see usability testing as the silver bullet to wipe out development problems. The U.S. National Institute for Standards and Technology has been effective in bringing together major software suppliers and purchasers to develop a Common Industry Format for reporting usability test results.[1]

The third method of user-centered design, *customer feedback,* comes into play once a product has been distributed. Software tools and network access have made a variety of monitoring and feedback tools possible. Simple counts of how many users open up a software tool and how often are very helpful. More detailed information of what features they use and don't use, accompanied by reports about what problems they run into, are even more helpful. Feedback from customer assistance professionals to designers about the problems users encounter is also very helpful. Interviews and focus groups elicit many suggestions for refinements and extensions. Detailed questionnaires on user interface satisfaction have been developed by several organizations. Innovative approaches to recording user actions and inviting user comments are likely to further accelerate improvements.

Although these methods have been emerging in the past twenty years, there is still a lag in adoption. In fact, there is resistance by some software engineers to adequately include user needs assessments and usability testing in their work. This results in high rates of failure in many projects, well described by the University of Colorado professor Tom Landauer in his thought-provoking book *The Trouble with Computers* (1995). Landauer makes the economic case for more attention to user interface design, which might have prevented numerous billion-dollar disasters. Even when projects succeed in turning out widely used products, there may be billion-hour fiascoes of user frustration.

Don Norman, a consultant and former University of California cognitive science professor, also promotes greater attention to user-centered design methods. In *The Invisible Computer* (1998), he encourages usability specialists to get more involved in early stages of product development and to become knowledgeable enough about market pressures to make realistic recommendations. He wants software engineers and technology developers to think more deeply about how human needs can lead to new ideas for products, such as Internet appliances and small portable devices.

Along with the new methods, many designers are adopting new metaphors of holistic understanding and planned development, often taken from architecture. Traditional architects combine client needs with technical realities and aesthetic styles, while keeping budgets and schedules realistic. In the best case, an architect meets with clients in their current living room to see and understand their lifestyle needs. They might want a personal home that combines space for children and an office for each parent. After talking about daily patterns, the client and architect might flip through books of sample homes to look at layouts and styles. Then the architect prepares a drawing for discussion with the clients. The right amounts of light, traffic flow in harmony with work and family, isolation of sound, and balance of private and open space are all important factors. Then, once these high-level choices have been made, the clients can choose the wood floors, carpeted stairways, and natural stone walls that all contribute to a successful project.

When the great American architect Frank Lloyd Wright (1867–1959) talked about his organic architecture, he made sure to include form following function. Successful buildings serve basic user needs. But he also recognized the importance of romance, tradition and ornament: "Poetry of form is as necessary to great architecture as foliage is to the tree" (Wright 1953). He was famous for making functional and aesthetic environments. He wanted to make beautiful buildings that worked for his clients.

This integration of functional and aesthetic aspects in a way that pleases users is very much in the spirit of Leonardo da Vinci. We admire his scientific spirit, his artistic skill, and his desire to produce something of value that also pleased his patrons. Leonardo skillfully played the lyre to satisfy Ludovico Sforza, the Duke of Milan, while exploring the mathematics of music and science of acoustics. He painted the mysterious *Mona Lisa* to please her husband, Francesco del Giocondo, while demonstrating his visual insights to facial anatomical details of her cheekbones, mouth, and eyes. Even the background shows off his knowledge of geology, plants, and river ecology.

Leonardo integrated art with science to serve a practical purpose. So for us, as technology users and designers, the take-away message is, technical excellence must be in harmony with user needs. Form should follow function, and then there is still plenty of opportunity for ornament. As designers we should strive to use appealing graphical designs that convey information effectively. As users we should celebrate designers who make products that are usable, useful, and enjoyable.

However, when you experience problems, you can promote better designs by speaking up. You should complain about a Web site that uses special plug-ins and large files with animated graphics to convey a simple message that would be best in plain text. Fancy graphics can be pleasing to some users, but the requirement for advanced software or broadband network access undermines the pursuit of universal usability. You should also complain to companies when e-commerce order forms are too long and confusing. The marketing manager may want to know everything about you, but disrespect for your time and privacy should make you angry enough to send e-mail.

MOORE'S LAW REEXAMINED

Even as our goals are changing we need to respect the past and learn from it. Five decades of amazing technological progress by devoted old-computing engineers has changed modern life and produced remarkable commercial successes. Faster computer chips and bigger hard drives have become predictable innovations for desktop and laptop computers. Moore's law, the doubling of computer power every eighteen months, has become the modern equivalent of Newton's laws of motion. The technological force of gravity keeps pulling computer prices down, even as their power increases. Higher bandwidth and more storage tempt the techno-savvy purchasers to buy yet another computer, but for many users these features and the focus on technology are losing their attraction.

The growing expectations of users are the propelling force behind technology evolution, and they are gravitating toward personal needs. The new computing is not about what computers can do, it is about what users can do. They want to communicate more rapidly with more people, learn about their health concerns in greater depth, and shop more conveniently. And when these basics are attended to, they want to accelerate their entrepreneurial dreams, pull together a family photo album, and plan their next vacation.

Similarly, professional users outside the computing industry are moving beyond first-generation tasks of managing mailing lists, keeping profit and loss spreadsheets, and computing satellite orbits. They see the new computing as a tool that enables them to form effective communities of practice, invent novel marketing partnerships, or form collaborations to discover environmental influences in atmospheric patterns. Administration shifts to innovation, supervision becomes inspiration, and data management gives way to knowledge management.

Thomas Friedman, the *New York Times* foreign affairs columnist and the author of *The Lexus and the Olive Tree* (2000), characterized similar transformations in world markets and diplomacy. He described the old days of the cold war as a time when the world was defined by competition, borders, and walls, especially the Berlin Wall. The new world of globalization is defined by collaboration and networked communications, especially the World Wide Web. He stresses the increased interconnectedness of markets and the empowering effect on individuals. Friedman claims that in the old world you were defined by your competitors, in the new world you are defined by your collaborators and partners.

The new computing is about collaboration and empowerment— individually, organizationally, and societally. The old measures of megahertz and gigabytes are less important. The new metrics start with how many messages you get per day, how many groups you contribute to, and how many other people link to your Web pages. These basic measures assess interaction, participation, and influence. Users are eager to interact with friends and family, participate in professional discussions, and influence their colleagues by presenting their viewpoints. Users are also enthusiastic to interact with like-minded users, participate in online communities, and influence political processes.

But empowerment is also about innovation and creativity; so the new computing users are counting the number of slide presentations they have presented, Web sites they have designed, and discussion groups they contributed to. Users are also proud of how many group projects they have participated in, such as writing for an online magazine, joining a neighborhood group, helping a start-up company. Most creative people also want to be recognized, noticed, and valued. So artists count their e-gallery visitors, musicians count how many times their songs are downloaded, and scientists count the references to their online journal papers.

Collaboration is influential at personal and organizational levels. More jobs require close collaboration with people who are often far from your desk because they have skills, contacts, or resources that you need to complete your work. Organizational collaboration is necessary for the same motivations. Even fierce competitors like Microsoft, Oracle, Sun, and Apple collaborate on industry standards panels and work to ensure interoperability of their products.

Organizational empowerment is exemplified by companies finding new markets on the Web, universities seeking distance-learning students, or museums providing public access to digitized copies of their art collections. Societal empowerment includes national projects to preserve cultural heritage, online access to legislative drafts to enlarge participation in democratic processes, and local language Web sites to support regional and ethnic groups.

This natural evolution from technological capacity to human capability parallels what happens in many fields. Early automobile designers had to concentrate on engine capacities, whereas modern automobile manufacturers understand the need for a holistic perspective on the users' transportation experience. The early measures were cubic centimeters of the automobile engine's displacement and its horsepower; modern advertisements promote the users' comfort, status, and safety. The measures for computing are also changing. Advertisements for portable personal devices no longer indicate chip speed or megabytes of storage; they focus on how many address book entries you can store.

It is a common aphorism that you are what you eat, but in the world of emerging technologies you become what you measure. As long as Web site managers only count the number of hits (pages requested) to their Web servers, they cannot easily understand user behaviors. However, e-commerce merchants are beginning to measure the efficacy and impact of Web sites by the conversion rate (what percent of visitors become purchasers) or the number of returning customers. Thinking about customer experiences is likely to lead to more effective Web site designs.

As long as school administrators only count how many classrooms are connected to the Internet or how many computers have been purchased, there will be little educational progress. Shifting to learner-centered measures such as how many e-mails were sent between students and teachers or how many Web pages students created is likely to steer educators in more productive directions. Then they will consider measuring student accomplishments by the number of links to student science projects, number of downloads of the school newspaper, or number of visitors to the environmental awareness Web site about the nearby wetlands marsh.

Users of information and computing systems are less interested in the size of their machine's RAM or their network's bandwidth than in finding a job listing easily and downloading the latest music video quickly. Most users want the hardware to be hidden under the hood; all they want to see is the dashboard—the user interface, with its colorful icons, familiar windows, and inviting buttons. Those who worry about Moore's Law are stuck on measuring the machine, but those who study users will learn that users simply want better experiences. The users want more information, better relationships, more chances to create, and better ways to send the world their message. The shift to user empowerment and collaboration are only two of three changes in the current evolutionary transformation toward the new computing.

FROM AI TO UI—ARTIFICIAL INTELLIGENCE TO USER INTERFACES

The third shift to new goals is the profound change in our perceptions of computing. Beginning in the 1940s the public's perception of computers was shaped by headlines about "giant electronic brains," then in the 1950s by "artificially intelligent" machines. Some early researchers, excessively influenced by the narrow goal of replicating human behavior, promised a General Problem Solver, medical diagnosis machines, and robotic house cleaners.

The mimicry game was played repeatedly as some technology developers sought to create machines that did what people did. This replacement strategy led to attempts to build computers that could see, hear, think, and even learn. Of course, there were success stories of computer vision systems that could detect manufacturing flaws or focus a camera lens, but emulating the richness of human perception has become a less significant goal. The mimicry game has faded, while practical applications have flourished. The explosive growth has been in visual image processing to manipulate still images, stored videos, or live Webcams. Image compression, manipulation, organization, and retrieval technologies are the big sellers, not image understanding.

Similarly, machines that learn remain an intellectual curiosity, but distance learning and interactive learning environments have become big business. Facilitating rather than replacing human performance is usually the winning strategy. Medical diagnosis by computer to replace physicians was the wrong goal. The success stories of the past decades have been support tools for physicians, such as improved medical imaging systems and remarkable medical lab tests based on an analysis of your DNA. Then your physician can make a more accurate diagnosis and work out a treatment plan for and with you (see chapter 7).

The quest to mimic human performance is misguided and largely counterproductive. Computers are not people, and people are not computers. Users usually want to be in control and resent the deception inherent in anthropomorphic designs that mimic human form and behavior. This idea of making computers resemble humans has a long history but gained adherents under the term *agents,* which were promoted by Nicholas Negroponte as early as 1970 (Negroponte 1995). However, users consistently rejected products developed as humanlike agents. Talking cars and soda dispensing machines came and quickly went in the early 1980s, along with humanlike bank teller machines. An early 1990s disaster was the $1.6 billion Postal Buddy, a life-size humanlike kiosk to provide postal services.

A grander failure was the social interface theory instantiated in Microsoft's BOB and heavily promoted by Bill Gates. Released in 1995, this $100 million fiasco was rejected by experts and novices because its cute conversational characters and distracting three-dimensional graphics interfered with work. But elements of BOB were revived with supposedly helpful characters such as Clippit in Microsoft's Office 2000 products. Clippit, the paper-clip character, would pop up and offer advice about writing letters or laying out your slides. It was supposed to be cute and likable, but Clippit was widely seen as an irritating intrusion and was finally removed by Microsoft in spring 2001. Designers are often attracted to creating humanlike machines, but most users don't want a relationship with their computer, they want control over it.

Programs that mimic or replace human capability strike me as a rather modest goal. I'd rather see tools that empower people by making them a thousand times as effective as an unaided human (chapter 11). A bulldozer makes the driver stronger than the strongest human, a gun makes the hunter a thousand times more deadly, and a camera makes the photographer more precise and rapid than the best artist. Microscopes, telescopes, and CAT scanners extend human abilities in remarkable ways. Mimicry of human form and behavior is useful for electronic toys, DisneyWorld's audio-animatronic performers, and crash-test dummies. Successful new tools and user interfaces rarely imitate people but rather empower people to carry out their desired tasks. Remember, the new computing is not about what computers can do; it is about what users want to do.

Hollywood contributed to the erroneous expectations of many people. Stanley Kubrick's compelling film *2001: A Space Odyssey* portrayed HAL, a sentient computer that was a very lifelike member of the spaceship's crew. HAL got many of the best lines and was often more emotional than the unsmiling human crew. The *Star Trek* television series reinforced beliefs that thinking and talking computers were close at hand. Talking machines are a great plot device and help keep the audience informed, but they set up the wrong expectations. Speech recognition technology is impressive, but the success story is the visual interface of the Web that gives users more information and rapid control.

Remnants of these artificial intelligence visions lingered on in the movie *A.I.* which portrays what could go wrong in building a robotic child that loves. Ironically, this movie, initiated by Stanley Kubrick and completed by Steven Spielberg, was released in 2001. However, it seems like an artifact of the old computing. Developers of the new computing know that it is not as useful to think about what computers can do as it is to think about what users can do. Even as progress has been made in machine translation of natural languages, speech recognition, and robotics, the measures of success and the attention of users are shifting.

The shift in attention is from AI to UI—from artificial intelligence to the user interface. The goal of UI is to develop designs that enable users to satisfy their genuine needs. The focus is not on AI but on U and I (you and I). A "smart" computer that takes family photos for you automatically is not as appealing as a digital camera that enables you to take photos of your baby's first visit with grandma. You can take pride in your photos only if you have control. A digital camera helps by relieving you of worries about focusing or lighting, and lets you see the photo immediately. If the photo is not appealing, there is still time to try again. If the photo is great, then you can easily annotate, store, and send it electronically to brothers and sisters. Successful designs amplify your photographic skills and enrich your family relationships.

The excess baggage from several decades of emphasis on what computers can do still burdens many technology developers. Letting go is not easy, especially because of cool demos and provocative fantasies about "expert systems" and "intelligent machines." But rule-based expert systems are no longer a major research topic and intelligent machines are a growing anachronism. "Smart homes" and "smart phones" are giving way to comprehensible products based on consistent, predictable, and controllable designs that genuinely serve the user's needs.

After almost two decades of incubation, user-centered design and usability engineering are gaining widespread prominence. These new professions demand multidisciplinary skills, sensitivity to users' needs, and tireless devotion to refinement. In the old days, designers were prohibited from talking to users or just didn't bother, but the new computing requires it.

GUIDELINES FOR USER-CENTERED DESIGNS

To speed this movement towards user-centered products, designers are eagerly learning about validated design guidelines. Web sites and popular books offer thoughtful rules about how to use color effectively, write instructions with clarity, and organize Web sites that are navigable.

The detailed design guidelines are important, but this book emphasizes the high-level goals, such as comprehensible user interfaces that generate feelings of mastery, satisfaction with accomplishment, and a sense of responsibility. This is far from the current state of confusion, frustration, and anxiety that most people experience when they use computers.

Future interfaces will be comprehensible to you because they are consistent, predictable, and controllable. Consistent designs use orderly layouts that align labels and group related items. They include appropriate color-coding to show relationships among components, provide warnings where needed, and highlight special cases. Consistent designs stick with meaningful terminology and present understandable sequences of instructions. Consistency is often invisible, because keeping the corporate logo in the upper left-hand corner or using the same color scheme seems natural. Only when the logo jumps to the right or the colors change do you notice.

Predictable designs enable you to gain familiarity and confidence in using your computers because you have a clear model of what will happen after each selection. You expect, after putting a book in an electronic shopping cart, that you can remove it or return to it a week later. Predictable designs apply meaningful metaphors, such as a shopping basket or e-mail in-box, and familiar conventions, such as the use of Save, Print, Open, and Close.

Controllable interfaces give you the power to do what you want. You can combine a photo with text to make a wedding invitation or copy part of a spreadsheet into an e-mail document. When you change your mind, you can revert to a previous version of your invitation or reformat the spreadsheet to provide input for a simulation program. Controllable interfaces are adaptable by users so that they can customize their screens and avoid unneeded features.

Day by day, users come to recognize better designs. They appreciate the prevention of error messages, they value a well-laid out Web page in an easily navigable Web site, and they cheer when they quickly find the product they want in a large catalog. E-business success stories are often tied to interface design improvements, as in the case of Southwest Airlines (figure 4.1), which has four times the industry rate of online ticket purchases because its compact Web pages enable rapid decisions (Hansell 2001).

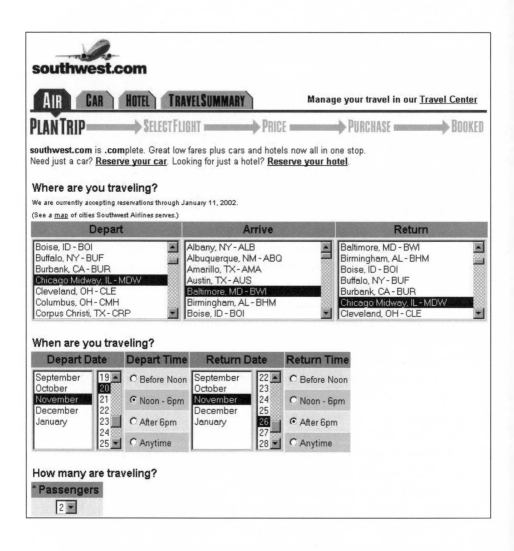

4.1   Southwest Airlines Web page (part), <http://www.southwestairlines.com/>,
"Reservations." Courtesy of Southwest Airlines Co.

Some design principles generate lively controversies. For example, many early Web site designers erred on the side of having too few links, thereby creating long paths to reach items in a product catalog. Such Web sites were difficult to explore because users had to go through too many steps and backtracking was tricky. After much discussion and careful experiments, designers generally agree that broader, shallower tree structures with more than one hundred links are more helpful to users. Designs with many links may result in "busier" web pages, but the payoff is dramatic in helping you to find your way to products, services, and information. This result was shown and replicated in a half dozen scientific studies that repeatedly demonstrated the benefits in faster task performance with broader, shallower tree structures on commercial Web sites. Yahoo! (figure 4.2) and eBay (figure 4.3) are great examples of the benefits of many links on the home page to provide an overview of what is available and reduce the number of steps it takes for you to get your job done.

Other design elements have also improved the user experience: faster system response time, larger screens, compact layouts, orderly alignment of fields, and much more. Scientific evaluations influence designers and developers of guidelines documents and software tools, which in turn provide the foundations for the next generation of user interfaces.

As books with design guidelines spread, new designers can learn from their peers, much as Leonardo's contemporaries shared best practices in their workshops. Guidelines are a good starting point, but the diversity of users and the range of technology must be addressed in order to reduce differences in Internet participation. One important goal is to push for universal usability, in which everyone can be a successful computer user. Universal usability makes good business sense because it creates larger audiences for commerce, entertainment, and education. Universal usability is vital to democratic principles because it generates an informed public, gives equal access to services, and encourages contact with government officials.

Dramatic differences in user participation were shown in a July 1999 report from the U.S. Department of Commerce titled *Falling Through the Net: Defining the Digital Divide* (NTIA 1999) and the follow-on reports called *Toward Digital Inclusion* (NTIA 2000) and *A Nation Online* (NTIA 2001). Well-educated people are still seven times as likely to be users of

Calendar   Messenger   Check Email   What's New   Personalize   Help

Yahoo! Messenger   Get your Web address now! YAHOO!   **Personal Email**
create your own webcam   CLICK HERE!   Domains   you@claim-your-name.com

[   ] [Search] advanced search

new!   **Yahoo! Fantasy Football** - season starts in 8 days!   new!

Shop Auctions · Autos · Classifieds · Shopping · Travel · Yellow Pgs · Maps **Media** **Finance**/Quotes · News · Sports · Weather
Connect **Careers** · Chat · Clubs · GeoCities · Greetings · **Mail** · Members · Messenger · Mobile · Personals · **People Search** · Photos
Personal Addr Book · Briefcase · Calendar · My Yahoo! · **PayDirect** **Fun** Games · Kids · **Movies** · Music · Radio · TV   **more...**

Find and Buy Anything on Yahoo!

Auctions	**Classifieds**	**Shopping**	**Features**
· Jeff Gordon	· Autos	· Apparel	· Back to School
· Barry Bonds	· Careers	· Books	· Store Builder
· IBM	· Real Estate	· Computers	· Auctions Booth
· Morgan Dollars	· Rentals	· Electronics	· Deals of the Week
· Longaberger	· Pets	· more dept	· Consumer Reports

Got Something to Sell? Auction it Now!

Arts & Humanities
Literature, Photography...

News & Media
Full Coverage, Newspapers, TV...

Business & Economy
B2B, Finance, Shopping, Jobs...

Recreation & Sports
Sports, Travel, Autos, Outdoors...

Computers & Internet
Internet, WWW, Software, Games...

Reference
Libraries, Dictionaries, Quotations...

Education
College and University, K-12...

Regional
Countries, Regions, US States...

Entertainment
Cool Links, Movies, Humor, Music...

Science
Animals, Astronomy, Engineering...

Government
Elections, Military, Law, Taxes...

Social Science
Archaeology, Economics, Languages...

Health
Medicine, Diseases, Drugs, Fitness...

Society & Culture
People, Environment, Religion...

In the News
· Tokyo explosion, fire kills 44
· Iowa slaying suspect charged
· Aaliyah mourned at NY funeral
· Little League pitcher overage;
  Bronx team's wins nullified
· U.S. Open - MLB - NFL
                                    more...

Marketplace
· Y! Shopping - Deals of the Week
· Y! Real Estate - rent an apartment
· Y! Careers - search IT Jobs
· Y! Autos - compare SUVs,
  sedans, sports cars

Broadcast Events
· 2:30pm ET Syracuse vs.
  Tennessee
· 6pm Miami Fla vs. Penn St.
· 7pm Wisconsin vs. Oregon
                                    more...

Inside Yahoo!
· Movies - Jeepers Creepers, O,
  American Pie 2, Rush Hour 2
· Pool - try our newest game!
· Play free Fantasy Football
· Make Yahoo! your home page
· Y! Games - spades, canasta,
  chess, dominoes, euchre, hearts

Local Yahoo!s
Europe : Denmark - France - Germany - Italy - Norway - Spain - Sweden - UK & Ireland
Asia Pacific : Asia - Australia & NZ - China - HK - India - Japan - Korea - Singapore - Taiwan
Americas : Argentina - Brazil - Canada - Chinese - Mexico - Spanish
U.S. Cities : Atlanta - Boston - Chicago - Dallas/FW - LA - NYC - SF Bay - Wash. DC - **more...**

4.2   Yahoo! Web page (part), <http://www.yahoo.com>. Reproduced with permission of
Yahoo! Inc. © 2000 by Yahoo! Inc. YAHOO! and YAHOO! Logo are trademarks of Yahoo! Inc.

4.3   eBay Web page (part), <http://www.ebay.com/>. These materials have been reproduced with the permission of eBay, Inc. Copyright © eBay, Inc. All rights reserved.

the Internet as poorly educated people. Similarly, wealthy households were three times as likely to be users of the Internet as poor households. Lower costs of computing and Internet access will help reduce this disparity, and improved designs will also help. Appropriate content and better training will also make Web services more useful to more people.

Designs should accommodate novices, non-English speakers, the disabled, the elderly, the anxious, and low-motivated users. Similarly, content should be adjusted to suit the needs of low-income users, diverse cultures, multiple ethnic groups, and minorities. Searching for a low-tech job or a babysitting cooperative in your neighborhood should be as easy as locating a book on Amazon.com.

Good design can also allow users with older machines, slower network connections, or smaller portable devices to accomplish their tasks. Flexible interfaces should allow for reformatting to fit different displays and elimination of superfluous graphics. Finally, good design should enable novices to gracefully learn what they need to become competent users. This may be the hardest challenge of all, so it will be one of the key challenges of the new computing.

WHY FOCUS ON HUMAN-COMPUTER INTERACTION?

More than a dozen research journals in human-computer interaction (HCI) are compiling practical results and theoretical frameworks to guide designers. Islands of predictive theory exist for well-studied problems such as keyboard data entry, pointing times for mice, and menu design. Innovative concepts, input devices, or visual displays are proposed at conferences around the world. Practitioners and researchers vie for attention from attendees and readers of proceedings. Professional societies in HCI have blossomed as independent groups, such as the Usability Professionals Association, and within existing bodies, such as the Association for Computing Machinery (ACM), whose Special Interest Group in Computer-Human Interaction (SIGCHI) holds an annual conference with three thousand attendees.[2]

HCI is also a growing academic discipline. Graduate degree programs in HCI are sprouting up in universities, and undergraduate curricula are being revised to include HCI. These developments are encouraging to those who promote the new computing, but defenders of the old computing do not give ground easily. Many traditional computer science departments resisted the new computing, causing their universities to sprout successful new units such as the Human-Computer Interaction Institute at Carnegie Mellon University or the Media Lab at the Massachusetts Institute of Technology.[3] Interdisciplinary groups, such as those at Stanford University or the University of Maryland, bring together faculty and students from multiple departments.[4] These success stories in human-computer interaction and user interface design are paralleled and emulated at other universities, but change often comes slowly.

The resistance comes from technology-centered researchers who value mathematical formalism more than psychological experimentation. The abstractions of an algorithm provide clear results, whereas the diversity and unpredictability of users are disturbingly messy. There are personality differences; computer scientists and information technology professionals have the highest degree of introversion of any profession studied. They prefer to work on problems in isolation, so the social issues of dealing with real users may be troubling.

Leonardo would have smiled in recognition about the struggle to gain acceptance for the new discipline of human-computer interaction. He had to fight for respectability for the arts, especially painting. In Leonardo's time, the ancient arts, derived from Plato and Aristotle, were divided into the Trivium (grammar, dialectic, and rhetoric) and the Quadrivium (geometry, arithmetic, astronomy, and music). Mathematics was the foundation for arithmetic and music, and measurement was the foundation for geometry and astronomy. Leonardo wanted to see painting accepted as universally as music and therefore set out to make a mathematical basis for painting that paralleled that of music. Burckhardt wrote about Leonardo's position in his classic 1883 book: "The art of painting, that is to say, the art of observing, was not only entitled to a seat among the Liberal Arts. She was the source and foundation of them all" (described in Richter 1969).

Leonardo was also in harmony with the empirical approach to life and to scientific research: "All sciences are vain and full of errors which are not born of experience . . . and not tested by experience" (Richter 1969). Leonardo wanted to try things out for himself, to build models, and to experiment with novel methods. I imagine that today he would be a supporter of building prototypes for novel user interfaces and doing usability testing.

THE SKEPTIC'S CORNER

Some technology devotees are probably angry by now. The argument for user-centered design reduces their central role and claim to fame. They sometimes respond that user-centered thinking is good, but that advances in technology must come first. They also feel that applications are not part of their focus and that technology progress should be unimpeded by additional constraints. Their perception is that user experience issues are the paint you put on when the building is done, but in the new computing user experience is the steel structure that frames the building.

I got exactly this kind of skepticism when questioning a promoter of "intelligent transportation systems." His goal was to increase capacity of highways by applying advanced vehicle sensors and automated control of cars. When I asked whether public transportation alternatives that might influence urban planning should be considered, he was adamant in rejecting this concern. His sole purpose was to increase road capacity. Even safety and cost were secondary for him. When asked about privacy concerns in tracking cars, he again argued that this issue was outside his concern and something for policymakers to decide.

One of the heroes of human-computer interaction, Don Norman, would also be skeptical of the capacity of usability professionals to change industrial methods. He faults professors who don't train students sufficiently in business practices and professionals who put usability above marketability. His critique has important lessons to teach, so I hope that the promoters of the new computing will learn them well.

Other skeptics question whether the new methods are sufficient to cope with the increased demands from complex software and multiple user communities. How can you test the millions of possible sequences of user actions? How can you do usability testing with varied user communities and the estimated one hundred and thirty distinct classes of disabled users? Usability testing has its limits, so other techniques such as automated testing and expert reviews need to be applied where appropriate. Perfection is not attainable, but improvements are possible.

Promoters of user-centered design should be aware of the resistance they may face. Copernicus and Galileo had a hard time gaining acceptance, too. The suggestion that the sun is at the center of the solar system was heretical. Similarly, the concept of user-centered design is difficult for some people to accept, but progress is clear day by day.

The Last Supper. From license-free "Leonardo da Vinci: Selected Works," Planet Art.

5

> UNDERSTANDING HUMAN ACTIVITIES
AND RELATIONSHIPS

*These notes reveal the intimate tie in Leonardo's thinking between . . .
phenomena in general and the need to put such information to practical use.*
—A. Richard Turner, *Inventing Leonardo* (1994), 184

WHY DO WE USE COMPUTERS?

Human nature and needs were not changed by the invention of computers. Human values endured even during the dramatic growth of information and communication technologies. People have always needed food, shelter, and medical care, and they always will.

People also need, seek, and thrive on emotional relationships with family, friends, neighbors, and colleagues. A nurturing parent and a caring friend will always be valued. Similarly, a neighbor's cheerful assistance and a colleague's supportive suggestion will enrich interdependence and strengthen trust. These human relationships grow when there is a shared history of positive experiences. The atmosphere of generalized reciprocity, the willingness to help others so that they will someday help you, smoothes the way for more ambitious cooperation and increases feelings of security.

When basic physical and emotional security are achieved, people then seek to establish commercial, political, and legal structures, and they contribute to these structures by their work and voluntary efforts. As civic structures are realized, people have growing amounts of leisure time and the security to get beyond basic needs. They can become creative in science, literature, music, and art, and they enjoy participating in entertainment, games, hobbies, and sports.

When these needs and desires are fulfilled, most people are satisfied. Having a good meal with your family and friends at a restaurant near your vacation home at a ski resort would make most people happy. But utopian images are only one side of human nature. Every positive can be linked to a potential negative: food can contain carcinogens, houses consume energy, and resorts can destroy the environment. And every happy relationship can fail: families can be torn apart, friends can become unscrupulous, neighbors can be deceptive, and colleagues can be competitive. Commercial firms can take advantage of customers, politicians can be corrupt, and legal systems can be manipulated. It can be a cruel world, but diligent planning can reduce the risks.

If technology developers start from an understanding of human needs, they are more likely to accelerate evolutionary development of useful technology. The payoff from a technology innovation is that it supports some human needs while minimizing the downside risks. Therefore, responsible analyses of technology opportunities will consider positive and negative outcomes, thus amplifying the potential benefits to society. These themes were inherent in the work of the social commentator and historian of technology Lewis Mumford (1895–1990), who characterized the goal of technology with quiet simplicity: "to serve human needs" (Mumford 1934). This straightforward phrase has been an inspiration for me, pushing me to construct principles to guide my own use of technology. Over the past year it has led me to a framework to help technology developers discover opportunities for innovation. Of course, the available technology greatly influences what one can build (Leonardo couldn't forge his giant bronze horse or make an airplane), but the ideas can still be valid. The available technology also shapes what one may consider possible—Leonardo never thought of CD players or cell phones.

As I studied my own use of computers and information technologies, I found it easy to interpret my usage in terms of satisfying my needs. I use computers to support my relationships with family and friends, to teach my students, to organize conferences with other professionals, and to buy books from online stores. My activities include gathering information, collaborating with colleagues, designing interfaces, and distributing my ideas.

When I interviewed other people about their activities, they also reported that gathering information, communicating with acquaintances and colleagues, and instant messaging to close friends and family were central to their daily use. These patterns were confirmed by a variety of user surveys by the U.S. Census Bureau, the Pew Center, UCLA, and others.[1] More than 80 percent of Internet users emphasize information gathering (especially medical, travel, and entertainment) and e-mail plus instant messaging. Most users of computers are not interested in the technology; they are focused on their own information needs and relationships.

Leonardo's thoughts about human needs are reflected in his list of four prime human activities: mirth, weeping, contention, and work. This was an appealing starting point, as he addressed emotional states that emerge in personal and business relationships. I also thought about Leonardo's integration of art and science in the service of practical purposes. The take-away message reverberated in my head: technical excellence must be in harmony with user needs. But what potent statement of human needs could guide design of information and communication technologies?

What better sources could there be than the Ten Commandments ("Thou shalt not kill, Thou shalt not steal, . . .") and the Golden Rule ("Do unto others as you would have them do unto you"). These are inspiring principles that should be in the mind of every user and designer, but I wanted more refined guidance that was easily translatable into technology innovation.

Another source was Thomas Jefferson's characterization of universal human needs in the U.S. Declaration of Independence, which proclaimed "life, liberty and the pursuit of happiness." I agreed that these are admirable goals, but I was looking for something that I could easily tie to new technologies. I registered Jefferson's goals in my mind and continued my search.

A later U.S. president, Franklin Delano Roosevelt, gave his Four Freedoms speech to the U.S. Congress (January 6, 1941) in which he looked "forward to a world founded upon four essential human freedoms." Roosevelt sought freedom of speech and expression, freedom of religion, freedom from want (economics and health), and freedom from fear (especially reduction in armaments). These also laid out useful ideas that are important to keep in mind, but I was after a more detailed link to activities and relationships.

In the 1950s, the psychologist Abraham Maslow proposed a hierarchy of human needs (figure 5.1).[2] He was writing at a time when psychoanalytic theory suggested that people were governed by subconscious motivations. He was also battling behaviorists' claims that people were merely stimulus-response machines and trying to shift researchers' attention away from abnormal behavior. Maslow spoke the refreshing language of "human potential" and described people as seeking creative expressions that enabled self-actualization. His early writings presented five levels of a hierarchy of human needs, building them up from the lowest level of survival needs to the highest level of fulfillment, which he called self-actualization:

1. Physiological: biological survival, food, water, air
2. Safety: secure house, no physical threats
3. Love, affection and belongingness: giving and receiving
4. Esteem: self-respect and respect for others, generates self-confidence
5. Self-actualization: fulfillment of what a person was "born to do" . . . "A musician must make music, an artist must paint, and a poet must write."

5.1   Maslow's hierarchy of human needs.

These are appealing because they deal with dangers to be avoided and goals to be sought. Levels 1 and 5 apply to individual needs, and levels 2, 3, and 4 describe relationships with others. This hierarchy appeals to me because it deals with an orderly spectrum of needs, is action-oriented, and focuses on relationships. I could begin to interpret how instant messaging or online communities support these needs. Maslow's hierarchy became an important guide to the new computing, but I was still eager to have a framework that made categories of relationships and precise activities more explicit.

My quest for clearer statements of human needs led me to a simple formula for life: living, loving, learning, leaving a legacy (Covey, Merrill, and Merrill 1994). Living and loving reiterate Maslow's levels 1–4, and learning and leaving a legacy expand on Maslow's level 5. The formula is brief and punchy, yet thoughtful and compelling. Covey and his colleagues suggest setting goals by writing a personal vision statement ("begin with the end in mind"), then choosing directions carefully. They push down to specifics, such as time management, empathic communication, and measuring progress. I also found that their descriptions of independence, dependence, and interdependence were effective in making explicit the multiple aspects of human relationships.

These philosophical statements are all helpful in shifting attention to universal human needs. They do set out important goals, but coming up with a framework to help technology developers seems to require an integrated activity-oriented approach that also identifies relationships— who does what with whom.

FOUR CIRCLES OF RELATIONSHIPS

In my quest to develop a framework for technology innovation I focused directly on growing circles of human relationships. In the old computing, computer usage was usually defined as a solitary experience, a concept that was encouraged by the term *personal computer*. But turning outward to focus on relationships led me to a fresh place where *family computer, corporate community,* or *civic network* might be appropriate terms.

At the center is still your personal use of the computer (figure 5.2). You may just want to listen to music, read the news, or write in your

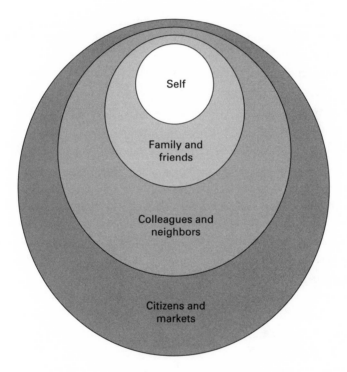

5.2   Circles of relationship.

diary. This is the private sphere, in which you are safe, secure, private, and free to create what you wish to satisfy your personal whims.

The second circle includes your enduring relationships with small numbers of trusted family or friends (2–50 people) with whom you have much shared knowledge and high expectations of meeting regularly. You've been cared for by your uncles and aunts, have played with your cousins and friends, and have gone to school with trusted classmates. They know a lot about you, and they would do a lot for you. You would trust them with your money, car, and emotions.

The third circle is much larger and includes the changing sets of professional colleagues or neighbors (50–5,000 people) who are moderately trusted, who share common interests, and who you expect to meet again. They might be members of your professional society or residents of your city, county, or state. You have a lower level of trust but share some common knowledge. You may not know other professionals or residents until introduced for a specific activity, but if they do the same work or live in the same neighborhood, you have lots to talk about.

Finally, the fourth and largest circle is defined by the citizens of a country or participants in a market (5,000 and more people). Your trust is low for them, your shared experiences may be few, and your economic situation may be different. Participants in a market such as eBay or MTV (music television) have some common ground, shared knowledge, and accepted social norms, but the risk of surprises is greater. People who regularly use eBay know about the latest policy changes and depend on the reputation manager to establish trust, but they are wary of each deal. Similarly, MTV fans share common knowledge of the latest music hits and fashions trends, but they don't leave their valuables in the open on the dance floor.

These four growing circles of relationships are characterized by size differences and the degree of interdependence, shared knowledge, and trust. It is an imperfect separation with fuzzy boundaries and boundary breakers, but it serves to identify current and potentially winning technology innovations. Buddy lists are for the intimate friends and family, whereas message boards and distribution lists work for colleagues and neighbors. Support for larger groups, such as novel Web strategies, are emerging to support relationships among millions of people who are citizens of a country or participants in a market. eBay, Nasdaq, and Amazon are examples of million-person online communities, although critics will argue about their boundedness and cohesion.[3]

Focusing on relationships is a new direction for many people in the computing field. After all, the basic notion of the personal computer was tied to the high degree of introversion among information-processing professionals. They usually prefer to be in their personal work space, and they believe that working alone is the fastest way to make progress, even if they could sometimes be more productive by cooperating with others.

It is not surprising that most software was designed for individual use, but as people with other personality types began using computers, their needs prompted the emergence of groupware and research on computer-supported collaborative work. As these new needs for cooperation appeared, new software and user interfaces were invented to provide appropriate communication. Of course, solitary work will always be necessary and group work has its problems. Many groups get into trouble, leading to spectacular controversies that trouble managers and participants. By contrast, individual failures tend to be more quietly covered over so that their damage is not noticed.

As a user, you might consider your balance of solitary and group work. Are there ways that you could use information and communication technologies to support your solitary work and to participate in the three larger circles of relationships? As a technology developer, could your innovation have multiple versions that are suited for solitary work or to support relationships in small, medium, and large groups?

As you shift your balance between solitary and group work, the benefits and dangers of each should be in your mind. Working alone frees you from interdependence but means you have only your own skills and knowledge to rely on. Working with others requires extra effort to build a trusting relationship, but you can share work and benefit from complementary skills and knowledge or simply split the effort to speed completion. Each approach has its satisfactions and frustrations, but drawing on both can provide the most productive and satisfying outcomes.

FOUR STAGES OF ACTIVITIES

The four circles of relationships are one dimension of my framework to accelerate technology innovations. A second dimension is needed to separate out the stages of activities that users participate in. One approach would be life cycle events such as birth, adolescence, marriage, and retirement. This generates new applications and Web sites, such as services to help new parents, teens in trouble, or wedding planners. A life cycle approach is helpful, but many important aspects of life happen between these memorable events. Another interesting approach is the rhythm of daily, weekly, or annual activities, so we'll save these for later refinements.

A better choice for an activity spectrum comes from studies of creativity (see chapter 10). The first step in a creative process, often called preparation, involves collecting information—just what today's Web supports quite well. In fact information technology has become the generic label for much of technology. But rather than focusing on how many gigabytes or pages of information are available (the old computing), we'll focus on the user's activity of collecting information (the new computing).

The responsibility for success in the information collection activity resides with the user. Users may *collect* information from family and friends, who are likely to offer what they know easily during a visit or phone call. The next circle of contacts includes colleagues and neighbors, who may be responsive to an e-mail message because of their expectation of reciprocity in their future information needs. Institutions such as professional societies, local government, corporations, universities, museums, and libraries offer informational Web sites that contain vast resources as part of their membership services, for a fee, or as a natural part of their institutional commitment. National resources such as the U.S. Library of Congress or the British Library, and marketplace databases such as eBay or Amazon serve broad audiences.[4] Users often have serious intentions related to shopping, but there's great fun in checking eBay to see the price grandma's crystal glassware is getting, or reading nasty and nice reviews on Amazon. Similarly, financial markets such as Nasdaq or the New York Stock Exchange generate vigorous enterprises, such as Fidelity, Smartmoney, and Charles Schwab, which provide voluminous information ready for collecting by consumers and professionals.[5]

Collecting information is the usual first stage of activity, and users may return to this activity repeatedly (figure 5.3). A second vital stage of activity involves relationships with others. The *relate* activity, consultation with peers or mentors, may occur in the early, middle, or late phases of a project. Relationships are so powerful an attraction that telegraphs,

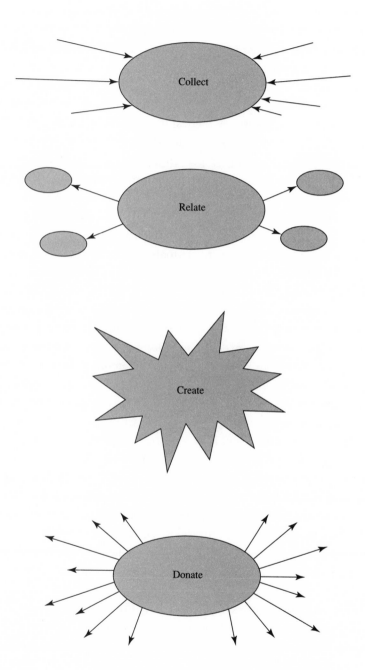

5.3   Four stages of human activities.

telephones, e-mail, and instant messaging grew rapidly and spread widely. Entrepreneurs quickly realized that the market was strong for communication technology. The thirst for relationships by phone and e-mail is so powerful that many people carry communication gadgets everywhere and spend large amounts of money for services.

By now, information (collect) and communication (relate) technologies are booming, so it is natural to ask what might be the next revolution? Leonardo now returns as the inspirational muse, because he reminds us of the strong urge to create. A broad interpretation of the *create* activity includes composing a song, planning a birthday party, launching a business, or organizing a social movement. So maybe the next revolution, following information technology and communication technology, is in innovation technology. This third revolution has already begun, and shrewd users and designers have already found the power inherent in creativity support tools. I'll have more to say later, in chapter 6, about the strong link between creativity and learning.

The fifth stage of human activity is what Maslow referred to as self-actualization and what Covey and colleagues called leaving a legacy. I'll use the term *donate* to complete the sequence with a rhyming scheme: collect-relate-create-donate. The *donate* activity covers giving to yourself, to your family, to your profession, to your community, or to your country. This is most visible in acts such as helping to care for friends and family, volunteering to help care for elders in a community center, or giving to national charities such as the Red Cross.

The concept of donation is also tied to dissemination of creative products. Composers of songs want to do more than write a great song, they often want distribution to and appreciation by others. Business leaders often talk about the desire to create value and develop successful businesses that change people's lives. Inventors want patents and royalties, while scientists want publications and citations. The desire for broad recognition and one's fifteen minutes of fame is widespread.

These four stages of human activity are not a complete representation of life, but they may help you by suggesting new ideas for accomplishing your goals. For example, you may think of shopping as merely finding the best price for your next car. But you could decompose shopping into collecting information about products, their features, and uses; forming a relationship with the seller; putting some thought into creating the deal; and then offering to be a positive reference for the seller.

This could make you a more effective shopper, and it might be more satisfying to you and the seller.

Similarly, you might rethink how you participate in sports or plan your next vacation in terms of learning more about the game or destination, building relationships along the way, creating something novel, and spreading the word about what you have done. The collect-relate-create-donate rhythm of activities may suggest new ways of thinking about old problems.

AN ACTIVITIES AND RELATIONSHIPS TABLE

You've probably guessed where this discussion is leading. The four circles of relationships combine neatly with the four stages of activities to make a two-dimensional grid: an activities and relationships table (ART) (table 5.1). This 4×4 table shows what activities you can accomplish with members of each circle of relationship using one of the technologies: information, communication, innovation, and dissemination. It's not perfect, but it may help you, as a computer user, to solve some of your problems in fresh ways. It may help technology developers in spotting new opportunities.

For example, if you are moving into a new neighborhood and need to find a doctor, your first step might be to collect information (first column of ART). Recommendations from family and friends would be a great starting point because you trust their comments. They know that you need a physician to take care of your two young children and a physician who is good in dealing with your asthma and your spouse's high blood pressure. You could ask your new neighbors, but you might be cautious since they may not know enough about your preferences or needs in medical care. You could try to track down listings of local physicians or local agencies that make recommendations, but you are likely to assume that these recommendations are biased or insufficiently tuned to your needs. Finally, you could consult national medical directories that list physicians by region and specialty, but this is generic information that is only a starting point for further inquiries.

Your second step would be to contact local doctors or healthcare organizations, describe your needs, and ask for references so you could

A R T	COLLECT Information >Read documents >Listen to stories >Explore libraries >Learn customs	RELATE Communication >Ask questions >Join meetings >Participate in dialogue and reciprocity >Develop trust and unity	CREATE Innovation >Compose, write, sketch, build, make >Brainstorm, visualize >Make plans and policies >Explore alternatives >Simulate outcomes	DONATE Dissemination >Write reports >Record history >Tell stories >Publish insights >Organize events >Advise, lead, care, train, mentor
Self				
Family and friends (2–50 intimates)				
Colleagues and neighbors (50–5,000 regular encounters)				
Citizens and markets (5,000+)				

TABLE 5.1   Activities and Relationships Table

communicate with current patients (second column of ART). References and testimonials can be an important part in building trust in new relationships. You might even contact medical review boards or state agencies to get details on performance of specialists or healthcare organizations.

If you've been especially diligent, you might organize all your information, ranking candidates by key attributes such as quality of care, cost, and convenience. If you were a truly devoted community activist, you could apply your creative energy to constructing a small handbook for newcomers to your neighborhood that records your information (third column of ART). Then you could help others by disseminating (donating) your information through a posting to a community Web site (fourth column of ART). The "donate" activity leads to creation of ever larger libraries of information that are the sources from which others can begin their collection process.

The new computing as mapped by an activities and relationship table could also be an opportunity for technology developers to invent software tools or services to organize and speed the process of choosing a physician. Spin-off ideas such as discussion groups among patients of specific doctors or among patients with specific diseases expand the possibilities. More ambitious opportunities to compare physician performance within regions or nationally become possible, as well as research to improve the quality of patient care.

This scenario portrays a diligent, motivated, and resourceful person who has the knowledge and time to find the right physician for his or her family. You might not be so dedicated, but you may find other ways to create a satisfying social community or to create something for your family or colleagues. Most people want to have friends invite them to go jogging and to live in a community where neighbors happily care for their children. Unfortunately, increased time pressures, longer commutes, higher expectations, and even growing use of the Internet can undermine your ability and willingness to give generously. But you can just as easily put greater emphasis on human needs and your community. You can use the new computing to restore the lost social capital.[6]

Many other everyday challenges that you face, such as finding a home to live in, starting a business, or getting a new job, could be facilitated by thinking about the circles of relationships and following the collect-relate-create-donate stages of activities.

For technology developers, the activities and relationships table might suggest new tools and services. The remainder of this chapter offers two case studies of interface innovations that were fostered by thinking of the activities and relationships table. The rows and columns are not sharply defined, and many activities will fit in several cells of this table. Imperfections and omissions are easy to find, but the goal of this table is provocative inspiration.

The first case study would have delighted Leonardo because it focuses on human needs in dealing with visual information, especially photos. The second case study covers desires for mobile and ubiquitous access, and suggests some novel ways and places in which to provide information, expand relationships, inspire creativity, and disseminate ideas.

THE EYES HAVE IT! VISUAL INFORMATION

It's a visual world! Images capture the thrills, emotions, and concerns of many people. Art can shock or inspire, concerned photography is a tradition, and family photos are some of our greatest treasures. Leonardo would have agreed. He wrote, "The eye . . . the window of the soul, is the principal means by which the central sense can most completely and abundantly appreciate the infinite works of nature." He believed that visual senses were the most important way to learn about our world.

It is not surprising that visual information is a vital component of the new computing. Most people depend on visual input for much of their understanding of the world around them and as a basis for further creative activities. Information visualization is becoming the next big success story as designers are succeeding in showing hundreds of times more information than tabular displays. These novel approaches show stock market trends, reveal disease patterns, or uncover flaws in manufacturing processes.

Mainstream success stories are inherent in the popularity of visual media such as photos, short videos, and clever animations. Digital photos and image-laden Web pages are already major computing applications, but users want still higher resolutions and faster downloads.

The possibilities for photo databases can be understood by using an activities and relationships table filled in with photo applications (table 5.2). The oldest applications are traditional database searches, such as

A R T	COLLECT Information	RELATE Communication	CREATE Innovation	DONATE Dissemination
Self	Digital photo import		Photo diary PhotoShop	
Family and friends	PhotoFinder PhotoMesa Family albums	Photo-sharing Web sites	StoryStarter Export to Web	Family photo histories
Colleagues and neighbors	PhotoFinder kiosk Corporate photo histories	Neighborhood photo sharings		
Citizens and markets	Library of Congress PictureQuest Corbis		PhotoQuilt	Web sites for photo exchanges

TABLE 5.2   Photo Applications in an Activities and Relationships Table

those presented by the Library of Congress. The Librarian of Congress James Billington has courageously promoted his vision through the National Digital Library Program to produce American Memory.[7] This system is making seven million objects available on the Web, organized into two hundred collections, many of which are already available, such as 1,100 Mathew Brady daguerrotypes of the Civil War, 25,000 architectural photos of Washington, D.C., from the 1940s and 1950s, George Washington's handwritten letters, Thomas Edison's films, and Walt Whitman's manuscripts. This is a national treasure accessible to all citizens and is structured as an information resource for teachers, students, journalists, researchers, history buffs, and others. As in a traditional and diligently catalogued library, images are annotated with information about the photographer, date, location, and a caption, all of which are searchable. You can search for "Missouri buffalo" or "Civil War general" to find photos with captions. It supports the "collect" activity in the broadest circle, citizens and markets. Other stock photo services such as PictureQuest or Corbis provide similar search services to broad markets.[8]

However, this is only one of sixteen cells in the activities and relationships table. Our research group took on the problem of organizing personal and family photo libraries, which addresses people's desire to collect and find photos of their families and friends (first and second rows, in the table). Personal photos deal with images of friends and family, typically at life cycle events (births, weddings, graduations) and on trips (whitewater rafting down the Colorado River, family visit to Disneyworld, summer trip to London). The distinctive nature of personal photos is that the same small circle of friends and family appear in most of the photos.

Personal photo images come from scanning services and digital cameras, which merely provide a sequencing number and then store the photos in directories. Few users are sufficiently motivated to give meaningful file names to each photo and organize them into named folders. But even those who invest the effort are limited by the crude file name searching that is available.

The challenge for users and designers is to overcome the paradox that family photos are among our most valued possessions but are rarely viewed. The value is exemplified by the story of a wedding that was restaged because film processors ruined the negatives. The entire wedding party and guests reassembled two weeks later, dressed in their fancy clothes to reenact the wedding for the cameras. Similarly, homeowners have run into burning or flooded homes to retrieve their photos. However, the paradox stems from the fact that most people cannot find their digital photos easily when they want them, and rapid browsing is not always possible. As users amass thousands of photos on their hard disks, they are stymied in trying to find that picture of grandma at Susie's first birthday party.

The personal photo paradox grows stronger over time because older photos are valued even more but viewed even less. It is just too hard to find that picture of grandpa in the navy during World War II. Therefore if search capabilities could be made effective, personal photo users might adopt them and be able to spend more time with their personal photos.

The main impediment to searching for photos is the complexity of specifying what you want and the difficulty of annotating the photos appropriately. If photos are annotated by date and place, as the Library of Congress does, then you can search for river photos of St. Louis, Missouri, or Cairo, Egypt. Places, dates, names, and nouns are relatively easy to search for, but images of concepts such as parenthood or themes such as mourning are more difficult to retrieve. Many thematic taxonomies or thesauruses have been built, but accepted standards are not in place.

Natural language techniques for expanding keyword searches can be helpful. For example, programs can use synonyms (*child, youngster, youth, kid, infant*) or hierarchical terms ("New England" is the more general term for "Maine, Massachusetts, New Hampshire, Vermont, Rhode Island, and Connecticut"). Some institutions, such as the anthropologically oriented Musée de l'Homme (Museum of Man) in Paris, organize photos by topics such as agriculture, housing, and religion, and the Detroit Institute of the Arts, for example, separates out still life, portrait, and abstract images.

Since annotation is so time-consuming, many researchers have attempted to use computer vision techniques to automatically analyze photos to recognize features, textures, faces, and colors. Color matching may help in finding sunset images, and feature detection can help find corners, lines, or circles, which do work in locating some images such as eyes or corporate logos. Photos of a person can be matched and located when good lighting conditions and face-forward poses are possible, but a general solution to the annotation and search problem is far into the future. Still, progress in automatic techniques may soon allow for counts of the number of faces in a photo or the recognition of indoor vs. outdoor photos. Other near-term possibilities are to distinguish buildings from trees, and male from female faces, which would be useful even if the results were not completely accurate.

An alternative to automatic analysis and annotation is to facilitate human annotation. While the Library of Congress and other institutions are willing to spend their resources to support detailed data entry, most users are not ready to spend hours typing in names of friends or family, locations, and contents. Even when they do, their inconsistent approach (Bill, Billy, William, Willy, or New York, NYC, New York City, NY) undermines successful searching.

Our approach to this problem with personal photo libraries is to enable you to drag names from a list of your family members onto the photos (Shneiderman and Kang 2000). This direct annotation interface, implemented in the PhotoFinder, records the names in a database that is easily searchable (figure 5.4). To make finding photos of grandma easy, for instance, we added drag-and-drop searching, so that you merely drag a name onto the search area and thumbnail photos of grandma will appear immediately. Our solution works well for personal photo libraries because even if there are thousands of photos, there will only be twenty to fifty people who reappear frequently.

Even before we finished the PhotoFinder, it was clear that finding photos ("collect" activity) was only part of personal photo library usage. Users wanted to send photos by e-mail to the people in the photos to confirm, capture, and relive their experiences. As we scanned older photos, users wanted to send photos to reminisce with participants about the event and tell stories to others who weren't there. The appeal of photos is that they are proof that something happened, the testimony that you did whitewater rafting through the Grand Canyon, caught the big fish in the South Pacific, or shook the President's hand in the Green Room at the White House. In terms of the activities and relationships table, PhotoFinder users were interested in more than just viewing the photos on their own; they wanted to use the photos to relate to other people.

And soon enough the retrieval of the photos became only the starting point for another creative activity. Users wanted to be able to export selected photos to a Web site and add commentaries to tell a story ("create" activity), such as the growth of a daughter from age one to age two. Users also wanted to print artistic compositions of multiple photos with

5.4   PhotoFinder software tool lets users organize, annotate, find, and share personal photos;
<http://www.cs.umd.edu/hcil/photolib/>.

textual captions. This led us to add the StoryStarter component to PhotoFinder. It allows you to export a set of photos with annotations and captions to a Web site in a convenient way and allows additional editing using commonly available editors.

Of course, the goal for most users was more than to make these creative products; they also wanted to disseminate them ("donate" activity) to family and friends. Each of these user needs led to expansions of the PhotoFinder beyond its original conception. These experiences helped confirm the utility of the activities and relationships table, but the next steps were to use the table to invent new applications.

In seeking photo applications for colleagues and neighbors, we came up with business uses, such as real estate agents' taking photos of houses for sale, insurance agents' photographing automobile damage, or surgeons' recording injuries and treatment outcomes. Such activities ranged from retrieving the photos (collect) to sharing them with colleagues for a consultation (relate), to making a report with recommendations (create), and publishing them on the Web for others to use (donate). Specialized applications to support creative effort become possible, such as retrospective searches by insurance agents to find all cases of failed bumpers or roll-over survival. Similarly, surgeons could annotate surgical procedures and then search for patterns over thousands of cases.

Our research plan expanded to include group processes for annotation. We scanned more than four thousand photos from twenty years of conferences in human-computer interaction and invited our colleagues to participate (relate) in providing annotations and captions. We installed seven PhotoFinder kiosks at the April 2001 ACM Conference on Computer-Human Interaction in Seattle. During three busy conference days, hundreds of visitors found their way to our booth in the back of the exhibition hall to browse photos and reminisce with friends. They added more than one thousand name annotations plus four hundred captions, and contributed twelve hundred new photos.

This process among colleagues contributed to a history (create) of this community of researchers and developers. By the end of 2001 we established a Web site with all the images, annotations, and captions to provide a public archive for this emerging discipline (donate). Of course, respecting the desires of individuals for privacy is a central concern in moving from private collections to public archives. Photographs of participants at a public conference are relatively uncontroversial, but personal photos that become public could produce sensitive issues and require model releases, especially if commercial gain is part of the goal. Responsiveness to requests for withdrawal of photos must be part of any public display.

One appealing application for families and friends is to build a personal history database of photos and stories that allows rapid finding of joyous events such as births and weddings or somber occasions such as illnesses and deaths. It seems likely that many people will keep more detailed electronic diaries with pictures, audio recordings, and videos. Software for indexing, organizing, and exploring these complex personal information forms are likely to be a large opportunity in coming years. Thousands of scanned or digital photos can be included and indexed by person name, date, location, and event. Software tools with user interfaces to allow viewing thousands of photos and zooming in on ones to be examined carefully are becoming possible on ordinary laptops.[9]

Since browsing has become so pleasant, you can imagine sitting with your grandparents, looking over old family photos while you capture their stories in audio format. Then you could tie in to one of the genealogical databases that allows easy creation of family histories. The graphic display of family trees and temporal presentation give you an overview of family evolution and the social context of the time. Along the way you might get distracted by the story of an aging relative whom your grandparents mention. The photos and captions describe his colorful life history of ambitious business ventures and dramatic travel adventures. You recognize one of his children as a distant cousin who once visited from Caracas, and make a note to contact him during your next business trip to Venezuela.

Neighborhoods and companies have histories, too. The ease of capturing events and stories, coupled with improved tools for organizing and presenting them, encourages some users to construct histories and mini-museums of their neighbors and colleagues. If Brooklyn is your home, you can find out about it, its neighborhoods like Prospect Park or its history, just by clicking to their Web sites.[10]

Similarly, IBM's ambitious archives are online and Intel's physical museum has a rich Web version.[11] Smaller companies can also have their history on the Web with photos of the founders and stories of how the company grew. The aging storytellers and devoted archivists of many minicultures become empowered and celebrated by their use of technology tools to capture experiences and encourage community.

Still more ambitious efforts are coming from Corbis, a for-profit company set up by Bill Gates. Corbis has been scanning photos and paintings from the world's great museums and archives with the intention of selling rights. But the image scans are only the starting point. Annotation, indexing, and search services are needed to make such collections accessible. It will be interesting to see how Corbis compares to the biggest library of them all—the U.S. Library of Congress.

Royalty and presidents have libraries of their archives with photos of their accomplishments, but in the future more people will create museums on the Web and slide shows about their lives and ancestors. Similarly, today only a few diligent specialists amass photo collections, such as the architectural historian, the birdwatcher, or the plant collector. But future experts and enthusiasts of many sorts will be able to develop and distribute their photo collections. They could do this as a hobby or a business, as an avocation or as a serious research effort. The communal aspects of group annotation, e-mail distribution of photos, and countless Web-based photo libraries are likely to stimulate a still higher degree of visual literacy in a wider range of people.

One of Kodak's efforts, PhotoQuilt, enabled thousands of people to create a joint photo archive.[12] Consumers sent in photos with captions that were assembled into a huge quilt-like image. Users could zoom in on a section and then click to get larger images with captions. This collaborative creation involved thousands of contributors and millions of viewers, but it is only the beginning. A natural next step would be to take the wildly successful concept of peer-to-peer sharing of music files and allow sharing of photos. Photo-sharing participants could easily access family photos or form collections from databases of historic photos locked away in the personal archives of cousins, uncles, aunts, and grandparents. Such photos could be tied to each family's genealogical database, allowing relatives to view each other's family photos.

Photos are a popular form of visual information, but there are other forms, such as trademarks, logos, cartoons, advertisements, paintings, and maps. In the early days of the World Wide Web, a University of Maryland computer science graduate student, who was an enthusiastic skier, approached me about his Web-based library of ski maps. He had diligently downloaded and scanned hundreds of ski area maps and arranged simple indexes to find resorts geographically, alphabetically, by difficulty, and so on. With some help from the university's Office of Technology Liaison, he managed to pay for his studies and support his young family when he sold the rights to an Internet service provider.

You may have your own examples of how you, your family, your company, your neighbors, or your college alumni created photo albums, stories, and archives. Photos are a fundamental technology, and they become compelling when they are used to support human needs for self-expression, collaboration, and creative endeavors.

MOBILITY AND UBIQUITY: PALMTOPS, FINGERTIPS, INFODOORS, WEBBUSHES

An activities and relationships table can be applied to the needs of users anywhere and anytime (table 5.3). The success of portable devices from the venerable transistor radio to the cherished Walkman cassette tape player to the amazing MP3 digital music players demonstrate the strong desire consumers have for music on the go. The same users who want big screens and impressive desktop machines in their offices and homes, also want small portable devices to carry with them everywhere. They want big screens for viewing maps and designing houses, and small screens for stock prices, weather reports, and flight information anywhere, anytime. Leonardo was just the same, sometimes working with wall-sized frescoes and large portraits but also using many small notepads. He was a scribbler, a doodler, who took his notepads everywhere.

Recognizing the desire for mobility, the designers of the 1990s started to produce small devices. The Palm (figure 5.5) and the Psion demonstrated that technophiles would use well-designed portable information tools—truly information at their fingertips. (Bergmann 2000). At the same time, exploding growth of cell phones exposed the intense desire for communications. By the early 2000s the combinations became irresistible: wireless communications for palm-sized devices and bigger displays for cell phones.

A R T	COLLECT Information	RELATE Communication	CREATE Innovation	DONATE Dissemination
Self	Flight info Weather		Diary	
Family and friends	Address lists	Find-a-Friend E-postcards	Music play lists	Family vacation histories
Colleagues and neighbors	InfoDoors Gather e-mail	Send-a-Link InfoDoors	E-guestbooks	
Citizens and markets	Stock quotes WebBushes	Click-n-pay	E-guidebooks	Web sites for exchanging tourist info

TABLE 5.3   Mobile and Ubiquitous Applications in an Activities and Relationships
Table

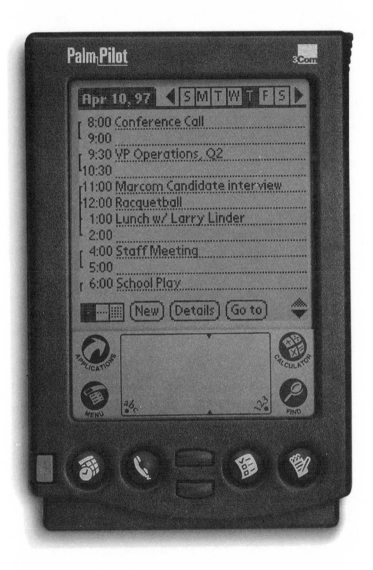

5.5   Palm™ hand-held computer. Palm™ is a trademark of Palm, Inc.

The Palm designers brilliantly focused on a few portable information needs: calendar, address book, to-do list, and notepad. The surprise was that users were willing to learn a variant of the English alphabet, called Graffiti (figure 5.6) so that they could enter data with a few quick strokes for each character. The surprise is even greater in light of the failure of the Apple Newton, which offered recognition for handwritten words. Apparently most users get greater satisfaction and utility from entering many easily recognized small strokes than a few often misrecognized handwritten words. Many users are willing to learn a new alphabet to obtain reliable data entry. Most users feel responsible for Graffiti stroke recognition errors, whereas they tended to blame the Newton for word recognition failures, This may be because the locality and cause of errors is clearer in Graffiti, making it easier to go back and get them right.

Palm add-ons, such as games and restaurant guides, appeared quickly, and now e-books and news headlines are growing applications as screen sizes grow and readability improves. In parallel, cell phone designers found a surprising willingness for users to enter short e-mails on the telephone keypad. Soon enough millions of young users shifted from talking to texting. Here again, mastering a new skill seems to have engaged many young users. The motto for the Yellow Pages directories was to "let your fingers do the walking," but with cell phone data entry, you let your fingers do the talking.

The Palm wins on screen readability, making it likely to grow as an information resource beyond news, calendars, or navigation. The cell phone wins on commercial viability because it is already tied to a pay-for-service mentality, making it natural to spend money by phone. Not only will one be able to buy stocks or airline tickets by cell phone but many little simplicities seem likely to appear. Why not pay for parking or a can of Coke by cell phone. As you park your car, just key in the code number pasted on the meter and a dollar gets charged to your phone bill. As you stand in front of a soda machine, just scan the code number pasted on the soda machine, and "You are on your way with Click-n-pay!"

But these devices are just the beginning of the miniaturization and pervasive implementation process. Wristwatch devices can already contain cameras or calendars, and other technologies will be embedded in

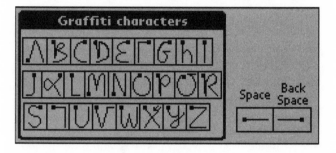

5.6   Graffiti® alphabet. Graffiti® is a registered trademark of
Palm, Inc.

shoes, bracelets, necklaces, rings, and clothing. Some will facilitate information gathering such as exchanging contact information, or be part of commercial activities such as payment processes, or collect medical information. Here again, visionary insights come from thinking more about human needs than technology possibilities.

The exchange of business cards is a delightful tradition that has acquired some appealing rituals, such as the formal Japanese offering process or the playful style of tossing them across a conference table. Palm users make a game out of their infrared beaming to send contact information. But imagine I want to gather the e-mail addresses of the fifty people at a meeting in order to send them a message. The Palm solution of beaming one contact at a time is much too slow and still makes it tedious to collect all the e-mail addresses into the header of a message. Recognizing the need for a GatherEmail tool is the first step, and then a dozen different technologies could be applied to solve the problem.

Other devices are likely to become fads. Lapel pins or earrings might become information exchange devices, with ten thousand names stored in shoulder pads or belt buckles. Other FingerTip applications would be to have rings, bracelets, or batons to operate devices around you. Imagine a ring that could be rotated to dim the lights of any room you were in, lower the air conditioning, or turn the sound volume up on the TV or radio. Maybe a bracelet that detected gestures would do the job, such as making a fist to turn the lights off or raising your finger from the stereo to the ceiling to raise the sound.

I've often wanted information services when traveling. Imagine that as you board an airplane you could check out if any people from your company or any alumni of your college were on board, or more practically, if anyone else was going to the Hilton Hotel so you could share the long taxi ride. These possibilities become viable as information access becomes widespread, but user control over privacy will become a growing concern. As I visit a new city, especially if I am wandering around on my own, I wish for a radar-like device—Find-a-Friend—that would let me identify friends who might be nearby to recommend a restaurant or show me around.

An extension of these ideas is that when attending a business presentation, you would automatically get the copies of the speaker's slides on your laptop—Slides-to-Go. This can be done by a wireless transmission (radio waves or infrared signal) of the file or simply its Web address— Send-a-Link. You could walk out of the meeting with the slides, the minutes, and the list of action items on your laptop or easily accessible from the Web.

But portable devices are only one manifestation of the intense desire for personal information and communication services. On the activities and relationships table, information tools that reach friends, neighbors, and colleagues might present opportunities for developers. Let's follow one more idea in depth.

Look at your office door at work. It probably has a nameplate in a wooden or metal frame, giving the room number plus your name and title. If you move, your nameplate gets replaced. Often your door acquires additional notes with schedule information, referrals for assistance, or a photo. Office doors often become a resting place for Post-it notes with messages such as "Out to lunch, be back at 2 p.m." or "Look for me in the kitchen." Travel plans such as "I'm in New York till Thursday" or "Vacationing in Paris till Labor Day" also wind up on many office doors.

Imagine mounting a Palm display on your office door with an Internet connection (wired or wireless). Voilà—the InfoDoor! The InfoDoor is an information appliance located at your doorway providing practical services such as personal scheduling, weather reports, or organizational announcements. It would have an Internet connection and a touch-screen surface mounted at eye level on or near the doorway to your office, linked to a server in the building. The InfoDoor would have an important role in emergencies, when it could steer you to safe exits in case of fires, toxic gases, or earthquakes. The lifesaving aspect of InfoDoors may justify their installation, but clever users will undoubtedly find other applications that are compelling or just plain fun, such as posting cartoons or personal photos.

A typical office building might have hundreds or thousands of Info-Doors. If bought in bulk and installed during construction, they would soon cost less than one hundred dollars per unit. The flexibility and openness for future growth should be appealing for "smart building" promoters. Normal operation would be quiet mode, in which the InfoDoor displays your name, title, or other standard information, but it could be changed to include a quote or joke of the day.

You could send a message to the InfoDoor to indicate that you are late for a meeting and you could provide information or instructions, for example, "I'll be in by 10:30 a.m., meanwhile please see Judy at the front desk." If you posted a schedule, then a visitor could select free times for a visit. If you were in a meeting, you could post a note saying, "Please do not disturb till noon" and encourage visitors to select a later time for a visit.

If you were not in your office or you were busy, the InfoDoor could provide referral information to appropriate providers, for instance, "For assistance till noon, see my secretary in room 472. To pick up a job application go to room 532. For information on available jobs, touch here."

You or the management could post announcements of time-sensitive events in the building, such as lectures, meetings, visitors, blood donation drives, charity programs, flu injections, and holiday gift sales. Other announcements could include weather reports, such as snow emergencies, hot weather warnings, and air-conditioning or heating changes. Traffic accidents, crime alerts, or early office closings could also be posted. Some of these can be sent by e-mail, but sending them to the InfoDoor would get them off e-mail and make them publicly available in familiar places.

Fire alarms or emergency messages could be sent with a warning tone. The messages could be more specific than current fire alarm systems and could direct people to the nearest safe exit while guiding fire fighters to the fire sources. InfoDoor alarms might satisfy emergency needs for rapid information during earthquakes, floods, toxic releases in industrial sites, explosions, or hostage situations in banks.

InfoDoors in office buildings, hotels, or homes are only one manifestation of ubiquity. Even in natural surroundings there are interesting opportunities to sprout new information, communication, innovation, and dissemination portals. I'll call these WebBushes. While every rock and tree might become the site for a new display device, let's explore a simpler approach of merely labeling distinctive objects with a barcode or small responder. This would make it possible to point your PalmPilot at a palm tree and find out what kind of palm tree it is, what its medicinal properties are, how it is used, and other bits of environmental, cultural, scientific, or historical information.

As you are rafting down the Colorado River, you pass some striking sedimentary rock face and click to find out what kind of geologic formation you are seeing, who first charted the upcoming rapids, and when the river crests in the spring. In-depth information about the location, its cultural importance, and tribal histories could all be available for the interested reader. Since each palm tree and river rapid would also have an associated Web site, as you travel your portable device would accumulate the sequence of URLs that defines your journey. Then when you return home, you could always retrace your steps because you would have a permanent record of where you were.

Text information is only a starting point for WebBushes. Each palm tree or river rapid could also be the basis for a photo database of professional photos in every season and through history. Visitors could leave their written experiences or photos for future reference or for others to see, possibly for a fee. They could dispatch e-postcards from memorable places and include photos from the here-and-now to connect with those who are far away.

Museum or hotel guest books are other applications that could be expanded in many tourist and natural locations. E-guestbooks could elicit user stories and encourage creative reporting that could enhance the experience for the teller and the recipient.

E-guidebooks are other opportunities for information collection and innovation. Think about riding along the Lewis & Clark trail from Missouri to Washington State, biking the 184-mile Chesapeake & Ohio Canal, or walking the Appalachian Trail from Georgia to Maine. At each rest area you could follow your interests and download onto your portable device the information about the next section of the trail. You could upload your sunset photos or add your observations of a rare heron. Some might argue that such media would distract from the natural experience, but they can intensify it as well by making visitors more aware of local birds, plants, or history. Not everyone wants to read the experiences of earlier visitors or leave their own comments, but many people seem to enjoy such exchanges.

Specialized information also spawns niche audiences. Guides for parents with kids, disabled tourists, amateur archaeologists, and so on extrapolate existing trends in guidebooks and other information sources Following the footsteps of frontiersman Davy Crockett or the residences of Leonardo may be specialized desires, but such individualized experiences give many people great satisfaction and a good story to tell when they get home. Registering at each destination and then having access to photos may be as much fun as visitors get at Disney's EPCOT when they stamp their passports at each country's pavilion.

The activities and relationships table (table 5.3) can now be partially filled in with these ideas. But these are just the beginning. By now you may have your own inventions and ideas for new products or services that could benefit yourself, your family, your colleagues, or wider circles of users. You may find new ways to contribute to the new computing, serving human needs for information, communication, innovation, or dissemination.

THE SKEPTIC'S CORNER

The activities and relationships table is not as neat as Mendeleyev's periodic table of chemical elements. Human activities and relationships are more fluid than puddles of mercury and harder to contain than clouds of hydrogen. The activities and relationships table is easy to complain about, incomplete, and too vague. But it does help shift the discussion from technology to human needs. It helps me think of whom I interact with and what I want to do in my life. It's not easy to make this shift in thinking, especially for those with technology-centered backgrounds, but putting user needs first is the key to the new computing.

You may be skeptical about some of my proposed photographic applications and even more doubtful about the InfoDoors or WebBushes. These are exploratory fantasies that may seem far-fetched, or they may send you to work writing a business plan to seek venture capital funding. If I've provoked you to do better, that will be an even happier outcome.

You may also challenge the fundamental idea that human needs should guide technology development. I've put this forward as a central thesis, even while I am aware of the enormous temptation, great power, and fun of thinking about technology first. Maybe I haven't won you over completely, but I hope that you will think about your and others' needs more often as you apply and design new technologies.

The next four chapters apply the activities and relationships table to e-learning, e-business, e-healthcare, and e-government. Those chapters explore these expanding applications from the perspective of the new computing. You will see opportunities for changing your family life and ways of working. You will see some of the dangers that lie ahead, but a cautionary approach might yield the most successful outcomes.

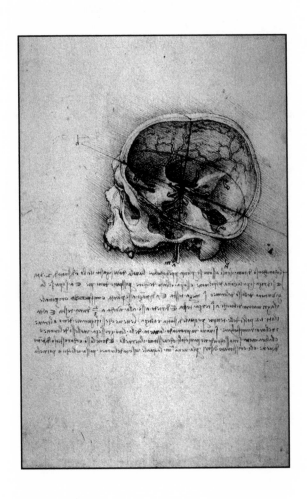

Anatomical Drawing of Skull in Profile to the Left. From license-free "Leonardo da Vinci: Selected Works," Planet Art.

6

> THE NEW EDUCATION: E-LEARNING

A system of general instruction, which shall reach every description of our citizens from the richest to the poorest, as it was the earliest, so will it be the latest of all the public concerns in which I shall permit myself to take an interest.
—Thomas Jefferson, letter to Joseph C. Cabell, 1818

WHY CAN'T EVERY STUDENT EARN AN A?

Memorable educational experiences are joyful and transformational. They enrich students with increased knowledge and skills, provide them with a satisfying sense of accomplishment, and reshape their expectations. In these compelling situations, students are driven by intense motivation that propels them to solve challenging problems and fills them with the thrill of accomplishment. They are proud of what they have done, have a clearer sense of who they are, and are ready to take greater responsibility for their education.

*L*eonardo learned his art by an active learning apprenticeship doing challenging real-world projects in the studio of Andrea del Verrocchio. Vasari's famous story of Leonardo's accomplishment said, "While he was studying art under Andrea del Verrocchio, the latter was painting a picture of S. John baptizing Christ, Leonardo worked upon an angel who was holding the clothes, and although he was so young, he managed it so well that Leonardo's angel was better than Andrea's figures, which was the cause of Andrea's never touching colors again, being angry that a boy should know more than he." Another version of this story ends with Leonardo's generous response; that it is the greatest compliment to the teacher that the student should exceed the master.

Verrocchio should get ample credit for having created an environment that engaged and transformed his students. They worked on individual projects and collaborations with each other or with Verrocchio himself. The art studio model has many advantages, but adapting it to the large number of students in modern schools is difficult. The standard lecture approach scales easily and some lectures are memorable, but studio-like challenges from teachers and interactions among students in small groups are usually more influential.[1]

Engaging experiences are often generated by an individual or team project that leads to a satisfying outcome. Enthusiasm is often high in class plays, orchestral performances, debate team tournaments, and science fair projects. This chapter proposes an active learning approach to education that integrates the new computing to create collaborative team experiences based on ambitious, authentic, service-oriented projects. New software tools facilitate such projects.

In reflecting on my experiences and my observations of students over twenty-five years, I have come to value student-led activities, such as class presentations, and team projects for an increasing share of my teaching. As I joined others who were also shifting from the proverbial "sage on the stage" to the "guide on the side" role, I developed a still greater appreciation of the opportunities for teaching and learning any subject in an environment that is rich in computing and communication technology.[2] My experience and examples are largely in a university setting, but variations on this philosophy are being applied to most settings and ages.

The old education emphasized acquiring facts and chunks of information that could be packaged into small teachable and testable units. Memorizing dates for Napoleon's rule, names of the U.S. presidents, or rivers of Africa is less relevant in an age of ubiquitous information. The new education accentuates critical thinking, analytic strategies, and working with people: family and friends, neighbors and colleagues, and citizens and markets. These goals are tied to improving communication skills and creative problem solving.

The old education emphasized competition, especially when students were graded on a curve. Since only a fraction of the students could earn an A, the need to beat your classmates and attract attention often dominated the need to learn. Students were prohibited from reading each other's work and required to work independently. The new education emphasizes collaboration, often requiring students to read each other's work. When the goals shift to improving the quality of every student's work, the grading must shift to allow every student to earn an A. This does not mean a lowering of standards, but a greater effort to motivate and guide every student to reach high levels of skill and knowledge.

To respond to these new education goals, teachers need a philosophy that they can adapt to their personal style, course contents, student population, and the new computing technologies. Based on the new computing's four activities we might rethink education in terms of collect-relate-create-donate:

COLLECT   Gather information and acquired resources
RELATE   Work in collaborative teams
CREATE   Develop ambitious projects
DONATE   Produce results that are meaningful to someone outside the classroom

The *collect* activity includes fact acquisition and traditional library research, but students need more guidance and tools to assess the validity and completeness of Web-based resources.

The *relate* activity encourages teachers to emphasize team efforts that develop communication, management, and social skills. The modern workplace demands proficiency in these skills, yet students are often forced to learn them on their own. Research on collaborative learning indicates that in the process of collaboration students clarify and verbalize their problems, thereby facilitating problem solution and anchoring, assimilating, and accommodating novel information. Collaboration has dangers, but when supported by the right tools, it generates intense motivation from many students, encourages learning from peers, and reduces dropout rates.

The *create* activity points to a fusion between learning and creative work. Student learning accelerates when teachers require creative outcomes in individual and team projects. In our modern world, learning seems useless unless it prepares students to be creative. Successful students create to learn, and learn to create. Software tools can make possible an unusually high level of creative accomplishment.

The *donate* activity stresses the benefits of having authentic, service-oriented projects that will be meaningful and useful to someone outside the classroom. Having an outside "client" further intensifies motivation, helps clarify goals, and provides training for future professional work. However, the right tools are needed for students to coordinate with their outside clients, who might be employers for students with part-time jobs or managers at volunteer organizations or campus groups. If possible, I would give a grade based on the amount of societal benefit produced during the semester.

Any definition of a teaching philosophy is merely a starting point for teachers, who adopt, adapt, and apply it in their own creative way. Variations of active learning have existed for thousands of years, as indicated by the ancient Chinese proverb

I hear and I forget
I see and I remember
I do and I understand

or by this quotation from Sophocles:

One must learn by doing the thing
For though you think you know it
You have no certainty until you try.

John Dewey (1916) developed the notion of authentic projects to
support education early in the twentieth century. His practical idealism
and devotion to empiricism would have been quite understandable to
Leonardo. Later, Jean Piaget's description of active learning and the cognitive stages of child development influenced many, including Seymour
Papert (1980). He and his disciples developed computer-based learning
environments for mathematics in which children wrote programs in a
powerful yet simple language called Logo. To make this environment
more concrete, he created a mechanical turtle with a pen that could be
programmed to draw squares, circles, and elaborate figures.

The case for active learning was boldly stated in 1971 by the Canadian educator Willard Wees in his aptly titled book *Nobody Can Teach
Anybody Anything:*

Whatever knowledge children gain they create themselves;
whatever character they develop they create themselves.

His strong statement about the necessity for students to create their own
learning experiences left a deep impression on me.

Wees's radical statement encouraged me to think boldly and ask,
Why shouldn't every student earn an A? The first response from educators habituated to competition is cynical: they assume that this goal is
attainable only by lowering standards. They still believe in grading on a
curve, where fixed percentages of students—say, 15 percent, 35 percent,
35 percent, and 15 percent—get A, B, C, or F grades. This strategy puts
students in competition with each other and often prohibits them from
discussing their projects with one another or getting help from others.

However, grading on a curve can encourage mediocrity instead of
excellence, and it often prevents students from learning communication
skills. By contrast, if instructors have a clear set of educational goals,
then isn't it possible to design courses so that every student attains
them? Shouldn't instructors, in part, feel responsible for every student's
success and therefore interpret a student's failure as their own? These
questions led me to teach by requiring students to critique each other's

projects and suggest improvements. Then students could learn from each other, and by reviewing other students' work, they would inevitably reflect on their own.

While I have not achieved the goal of having every student get an A, the dropout rate has declined even as the workload and my expectations have grown. I've come to see that the sound of learning is not my voice lecturing but the buzz of team discussions during a collaborative exercise. I've come to appreciate that the transfer of knowledge is not in my handouts but in the e-mail exchanges and instant messages among students.

There are problems. Two out of ten teams get into some trouble, and one in ten may have a serious conflict, but resolving such problems is part of what they are learning, too. I remember one team with Turkish, Israeli, and Jordanian students who learned to cooperate across languages, cultures, and conflicts. They learned about differences in expectations about sharing work and timeliness. The happy outcomes were that they produced a project that was published in a professional journal and that two of the three have remained in contact years later.

Information and communication technologies facilitate active learning and collaborative teaching methods. Students can create remarkable products and coordinate their work much more easily using these technologies. Taking inspiration from Leonardo, we might envision an educational Web tool, called LEON for learning online, that propagates the spirit of Verrocchio's studio. The home page for LEON would include appealing artwork and soothing music. Students and teachers could work together on ambitious projects that they post on the Web for others to review. I discuss the design of LEON later.

I am not alone in emphasizing collaboration, inquiry-based projects, and active learning. The American Association for Higher Education (AAHE 1987) has spelled out seven principles for good practice in undergraduate education, which include active learning:

> Encourage student-faculty contact.
> Encourage cooperation among students.
> Encourage active learning.
> Give prompt feedback.
> Emphasize time on task.
> Communicate high expectations.
> Respect diverse talents and ways of learning.

Involvement in Learning, the final report of the Study Group on the Conditions for Excellence in American Higher Education (NIE 1984), stated that "active modes of teaching require that students be inquirers-creators as well as receivers of knowledge." It called for involving students in faculty research projects, internships, small discussion groups, in-class presentations and debates, individual learning projects, and developing simulations, and stressed the importance of projecting high expectations and giving feedback.

The National Academy of Sciences (NAS/NRC 1996) emphasized student initiative: "Learning is something students do, not something that is done to them (20); . . . inquiry into authentic questions generated from student experiences is the central strategy for teaching science" (31).

Report to the President on the Use of Technology to Strengthen K–12 Education in the United States (PCAST 1997) took an even stronger position in supporting active learning and authentic projects: "Basic skills are learned not in isolation, but in the course of undertaking (often on a collaborative basis) higher-level 'real-world' tasks. . . . The student assumes a central role as the active architect of his or her own knowledge and skills, rather than passively absorbing information proffered by the teacher." Similar comments have been made in Britain (Dearing 1997; Hazemi et al. 1998).

For elementary schools, my colleague Allison Druin pushes these ideas still further by creating intense experiences in which children aged 6–13 become technology design partners (Druin 1999; Druin et al. 1999; Druin and Hendler 2000). Her goals involve technology innovation by her intergenerational teams, and the educational experience from doing authentic projects is powerful. Her kids even become co-authors of professional papers in respected conferences and books.

TEACHING AND TECHNOLOGY

No breakthrough in instructional technology will solve the education problem. However, the transformation to the new computing in education will bring positive changes as teachers integrate general computing tools such as word processors, Web browsers, e-mail, online communities, digital libraries, and simulations. There is already a lively competition between suppliers of educational software to manage courses and content, such as WebCT and Blackboard, that have many components that we'll want in LEON.[3]

However, educational technologies like LEON usually provoke controversy. Even books and paper faced resistance because they reduced human memory skills. Opponents might have argued that they would undermine the role of teachers. Proponents probably argued that textbook-assisted instruction meant that while students could learn on their own, they would still benefit from teachers, who could now concentrate on challenging, guiding, and evaluating student performance. Paper and books changed the content of education because memorizing lists of English royalty or medicinal plants gave way to consulting extensive royal genealogies or detailed pharmacological tables. Powerful technologies change our expectations and curricula.

Paper has an even more potent role than as a storehouse of knowledge. It achieves remarkable power when it is a blank sheet, inviting student creativity. But the transformational insight that students should be more than copiers took centuries to emerge, and must have been difficult to promote. By now, teachers assume that students should write as well as read.

Listening to radio and CDs and watching TV and videotapes have been proposed as educational panaceas, but they too have become only multimedia components for WebCT or Blackboard. These are largely passive media, offering limited capability for students to be creative, unless educators shift their focus to student content generation using these media. Once again, it is taking decades for educators to recognize that the most potent use of videotapes happens when teachers offers blank ones to students.

Similarly, the World Wide Web or LEON cannot be a solution to educational needs unless the creative component is included. We have to do more than teach kids to surf the Net; we have to teach them to make waves. Finding Web resources is fine; creating new ones is the key to the new education.

APPLYING COLLECT-RELATE-CREATE-DONATE

The vision for LEON is guided by an activities and relationships table adapted for e-learning (table 6.1). The Self row remains intact because students still need to work on their own for much of the time. The

A R T	COLLECT Information	RELATE Communication	CREATE Innovation	DONATE Dissemination
Self			Solve problem	
Team members and teachers	Information resources	Ideas, data, plans	Identify problem Brainstorm Refine solution	
Classmates and teacher aides	Information resources	Assistance	Critiques	Drafts
Clients and readers	Information resources		Apply solution	Final reports

TABLE 6.1   Education-Oriented Roles in an Activities and Relationships Table

Family and Friends row becomes the Team Members and Teachers, who are close contacts for each student. Student relationships with their teachers and other students are dramatically facilitated by e-mail. Student teams no longer have to find meeting times but can send ideas, data, and drafts to each other at any time.

The Neighbors and Colleagues row becomes Classmates and Teacher Aides, to whom a student might broadcast a request for help or information at any time. Finally, the last row of the table becomes Clients and Readers, who are people outside the classroom who benefit from the projects that could be posted on the Web for anyone to read. With these revisions to the activities and relationships table, we are ready to take a look at the activities: collect-relate-create-donate.

Collect: Gather Information and Acquire Resources

Studying the past is a useful guide to the future. Learning what has been done is a solid foundation for innovation. LEON would provide a student-oriented view of the remarkable resources on the World Wide Web and alert students to the dangers. It would index review articles and primary sources so that students could read the letters of George Washington without traveling to the Library of Congress. Students could use current environmental data to make their own measurements of deforestation and form their own theories of how to control it.

LEON would also guide students to recognize the difference between a drug company's assessment of the efficacy of its medications and clinical trials monitored by the Food and Drug Administration. Similarly, students need to learn to analyze news reports on current events or historical reviews based on the sources. LEON would have to help students learn that the Web can spread misinformation, rumor, and hatred just as fast and easily as authoritative reports.

Relate: Work in Collaborative Teams

Extensive descriptions of collaborative teaching methods reveal the rich set of possibilities available to teachers.[4] Collaborations can be as simple as a two-minute in-class exercise involving pairs of students or as elaborate as a two-year project-oriented curriculum involving dozens of students. Collaborative team projects have the potential to raise motivation, reduce dropout rates, and develop job-related skills, but collaborative methods are not commonly used. Most teachers have little experience with collaborative methods and team management because they were likely trained by lecture methods. Lectures can be effective, and technology can help improve them with presentation software and online demonstrations, but more active teaching strategies with collaborative teams offer appealing possibilities.

LEON would provide teachers with an array of collaborative methods and then guide students in effective participation. Term-length or shorter team projects done outside the classroom seem more acceptable to teachers, maybe because they minimally disrupt a lecture-oriented course plan. Students who work together on substantial team projects learn a great deal about project management, leadership styles, and efficient use of time. LEON would guide students in making a modular work plan, setting schedules, reviewing each other's work, and resolving differences. College-level collaborative projects are typically organized around half-hour meetings, separated by two to ten hours of individual work. Short presentations by project teams to the full class are an appealing way to develop students' public speaking skills. LEON would contain templates for talks and archives of exemplary student slide shows.

Computer-rich networked classrooms make possible a variety of collaborations. Students can create on their computers and show the whole class for discussion with a large-screen projector or by copying to every student's computer. By rapidly reviewing student work, everyone can see the range of good to bad work and can sharpen their reviewing skills. With appropriate software, classroom brainstorming among dozens of students can, within minutes, produce an amazing variety of comments, which can then be discussed or saved for review. Anonymous inputs encourage diverse and creative suggestions. Rapid class voting on the results of a brainstorming session typically leads to a spirited discussion (Alavi 1994). LEON would build on tools such as GroupSystems that allow group brainstorming, voting, and convergence through the Web or in electronic classrooms.[5]

For in-class collaborations, pairs of students rather than one or three per computer produce superior educational outcomes.[6] Pairs of students sit together verbalizing potential solutions; typically, one student deals with the computer while the other focuses on the problem. Regularly switching roles ensures balanced learning. As students explain what they don't know to each other, they solidify their knowledge and learn rapidly. Asking a good question is one of the golden keys to learning. Educational psychologists talk about meta-cognitive skills: the capacity of students to reflect on what they know and what they don't know. High-performing students constantly monitor their progress and bravely declare, "I don't understand," to their instructors or peers.

LEON would include e-mail and instant messaging—their low cost plus high payoff should make them useful in almost every course. LEON would also include listservs or threaded discussions automatically for every course, even before the first face-to-face meeting. These tools are wonderful for sending hints on assignments to an entire class or messages to individual students commenting on their work, especially when the next class is three or four days away. Students can also initiate questions at any time to the entire class, to individual students, or to the teacher.

This seven-day, twenty-four-hour classroom approach has the potential for creating an intense environment that may be overwhelming to some students and teachers. I have learned to advise students about how to manage their time and to set expectations of what level of effort I expect. My design of LEON would include monitoring tools that enable teachers to see which students have been overly active or inactive.

Experience is accumulating in many disciplines and educational settings, but many questions remain open: What size team is optimal? Team memberships assigned by teachers have been demonstrated to be more effective than allowing students to stick with their friends. But on what basis should teachers construct teams? Are uniform skill and motivation helpful? How does a teacher intervene when a team member consistently fails to perform? Should grades be assigned individually or to the team?

Research on collaborative teaching and learning continues with the promise of deeper understanding and refined guidance for teachers. Experience with LEON's collaborative software should lead to improvements for managing and guiding students, even for larger teams and larger courses.

Create: Develop Ambitious Projects

Teachers regularly wrestle with the formulation of team projects and in-class collaborations, especially since textbooks provide only modest help.

LEON's examples from thousands of courses could help. A natural project for students is to produce an online textbook or encyclopedia for their course. With a class of ten to one hundred students, this becomes a major effort, with an editorial board to develop an outline, specify an audience, produce a style guide, manage assignments, and arrange internal reviews. Twenty-four students in my graduate seminar on virtual reality produced a wonderful resource—Encyclopedia of Virtual Environments (EVE), which continues to be maintained by the University of Washington's Human Interface Technology Lab (figure 6.1).[7] Eighteen students produced a helpful guide to universal usability resources for Web developers (figure 6.2).[8] The idea has been repeated at the undergraduate level and at the elementary school level, a fifth-grade class produced a database on the animals of Africa for third graders.

The Web enables students to publish their projects, making them available for anyone. Students are anxious about making their work so visible, but it does push them to polish their projects more than in the past. The student critiques of each other's projects by e-mail to the authors and to me is an important part of the process. I require one paragraph describing what they liked about the project and one paragraph making constructive suggestions for improving the online presentation. I grade these critiques and give the students a few days to make changes before I announce the Web site (figure 6.3) to several newsgroups and listservs.[9] Students are proud of their accomplishments, and each has an impressive product to include in a job interview portfolio. Revised versions of several papers have been published in professional journals, conference proceedings, and online magazines.

I would want LEON to help me manage the critiquing process, making sure that every project was reviewed and that every student did a review. Then I'd want to monitor the changes made to each online report, currently an impossible task for me.

Donate: Produce Results That Are Meaningful to Someone Outside the Classroom

The rewards of helping others are especially sweet when you are also helping yourself. This can be the case when students work on service-oriented authentic projects for clients outside the classroom. Students

The Encyclopedia of Virtual Environments

Produced by the students of Dr. Ben Shneiderman's CMSC 828S Virtual Reality and Telepresence Course, Fall 1993

See also Ben Shneiderman's essay on *Education by Engagement and Construction: Can Distance Learning be Better than Face-to-Face?* for a discussion on the theory and background of this distance education project.

Abstract:

The Encyclopedia of Virtual Environments is an attempt to describe the technologies and techniques being used to produce Virtual Reality (VR) applications. It is a collaborative effort, whose contents and structure evolved through discussions that took place mainly over electronic mail, supplemented by some face-to-face discussion in class.

The list of authors contains biographic information and photos.

Be sure to look in the definitions article if you encounter any unfamiliar terms in the other articles.

Table Of Contents

1. **System Components**
   - A. **Sight**
     - Display Technologies
     - Graphics Subsystems
     - Rendering and Animation
   - B. **Sound**
     - 3D Sound
     - Voice Synthesis
     - Sound Sampling

6.1   The Encyclopedia of Virtual Environments Web page (part), <http://www.hitl.washington.edu/scivw/EVE/>.

 **niversal sability in ractice**

Principles and strategies for practitioners designing universally usable sites

Users with Disabilities	Special User Groups	Technology	Tutorial methods
Blind and low vision users	Children	Users with slow connections	Designs to help novice web users
Color vision confusion	Elderly	Users with screens less than 640 x 480	Online help design, email help methods and customer service guidelines
Cognitively disabled	Users with low education, low motivation	Telephone based access to the web (WAP)	
Deaf & hearing impaired	Users of other languages than English	Telephone based access to the web (speech recognition)	
Mobility impaired	Users from other cultures than the US		
	Cross language information retrieval	Textual equivalents for audio/video representations of content	

The goal of universal usability is to enable the widest range of users to benefit from web services. This website contains recommendations and information resources for web developers who wish to accommodate users with slow modems, small screens, text-only, and wireless devices. It deals with content design issues such as translation to other languages, plus access for novice, low educated and low motivated users, children and elders. The website also covers design guidance for blind, deaf, cognitively impaired, and physically disabled users. Each article has practical guidelines, web site examples, links to organizations, and a bibliography. For related information see www.universalusability.org and the information from the ACM Conference on Universal Usability (November 2000).

This website is a class project for Human Factors in Computer and Information Systems (Computer Science 838S) (Spring of 2001). It is a continuation of the UUGuide project started by graduate students in the Spring 2000 class. The courses were led by Prof. Ben Shneiderman Founding Director of the University of Maryland Human-Computer Interaction Lab.

Editorial board: Irina Ceaparu and Dina Demner
Last updated 5/17/2001

Related Links Universal Usability Template Privacy Policy Universal Usability Statement

6.2 Universal Usability in Practice Web page (part), <http://www.otal.umd.edu/uupractice/>.

STUDENT HCI ONLINE RESEARCH EXPERIMENTS

During the Spring 2001 semester, 12 student teams conducted empirical studies of user interfaces as their term-length project in Computer Science 434/838: Human Factors in Computer and Information Systems taught by Dr. Ben Shneiderman, Director, Human-Computer Interaction Lab. The online reports of these projects include links to previous work and related systems. Project webpages were created by each team, using this template. Student experimental projects are also available from previous years: 1997, 1998, 1999, and 2000.

Handheld Devices

A Comparison of Grafitti vs. the On-Screen Keyboard for Experienced Palm Users

Data Input Into Mobile phones: T9 or Keypad?

Which is Faster and More Accurate on a Handheld: Graffiti or Keyboard Tapping?

Web

"In Web We Trust": Establishing Strategic Trust Among Online Customers

Navigation Bars for Hierarchical Web Sites

The Menu Design and Navigational Efficiency of the E-Maryland Portal

Searching for Airline Tickets: A Comparison of Tabular and Graphical Presentations

Layout and Readability

Cross Language Information Retrieval: Layout Strategies for Gloss Translation

The "Degree Navigator" Nightmare: Taming An Overly Graphical User Interface

The Impact of Window Desktop Design on User Performance: Microsoft Windows Explorer vs. ClockWise Win3D

Visualization of Shallow Trees with Nodal Attributes using Fisheye

6.3   Student Human-Computer Interaction Online Research Experiments (SHORE) Web page (part), <http://www.otal.umd.edu/SHORE2001/>.

have skills that can be highly beneficial to many organizations, and students benefit by having a client whose needs promote advanced learning.[10]

LEON would provide an archive of previous student projects and structured processes for carrying out projects. My students have worked on campus-related projects, such as scheduling systems for the bus service and a television station, record keeping for a scuba club, a student ride board, carpooling, and an accounting system for the physics department. Off campus software projects have included donor and volunteer list management for a major charity, scheduling for a county recreation office, and information management for a day care center. Other students' projects have developed a guide to science education software for parents of junior high school children, a hypermedia guide to computer viruses, and a plan for computer usage in a local high school.

One of my favorite projects focused on helping elderly residents in a nearby retirement home. The students read the literature to learn what was known about elders' learning and using software products. Then they made several visits to the retirement home to try different training strategies. Their final report, citing the literature and relevant software, was written with recommendations for the director of the retirement home.

Sometimes I receive requests for my students to work on specific projects, but most students find their own projects. Their sources include their part-time jobs, hobbies, or student organizations and often their parents or siblings. Naturally, my students find computer-related projects, but faculty in archaeology, journalism, business, biology, and other areas have reported success in finding service-oriented authentic projects.

Clients' expectations should be kept modest and students must be recognized as volunteers in an academic setting. In my experience the clients have been satisfied with the experience because it gave them new ideas and often a prototype for a future project. One or two in ten projects result in operational systems that are used or serve as the basis for an immediate development effort.

The State of Maryland requires seventy-five hours of community service for high school graduation. I think this noble requirement should serve as a model and inspire a similar plan at the college level. Students are highly motivated by having a real client, and they are proud of their accomplishments. They also have an ambitious project in their portfolio when they go on job interviews.

While many school teachers have adopted classroom discussions, small-group activities, and authentic projects, these methods are less common in higher education. College teachers and administrators may still resist collaborative methods because of the novelty for teachers, concerns about adequate coverage of the required curriculum, and difficulties in giving individual grades. The image of the college professor as the "sage on the stage" is strong, so it may be initially uncomfortable to be the "guide on the side." When I first assigned an in-class discussion topic for pairs to work on for three minutes, I felt uncertain about what I was to do. But as the buzz of spirited conversation grew louder and was difficult to suppress, I began to appreciate the power of collaborative methods.

Teachers are expected to "cover the material" in their lectures, and they are gratified by the process. However, the result may be merely that they obscure the topic. Lecturers have little assurance that students are learning anything. Students should be uncovering or discovering the material for themselves. With in-class presentations and collaborations, student engagement is at least more visible. Traditional tests and individual homework assignments can be merged happily with occasional or regular collaborative methods to enable teachers to assess individual student skills.

Most students readily take to collaborative methods, but there are often one or two students in a class who request to work alone. They allege heavy work or family burdens, but ironically they often argue that they would be willing to take on the responsibility of doing the entire project on their own. Computer science students (and professionals) rank high on introversion scales, so their resistance is understandable. I do require team participation and indicate the importance of learning to work with others as a natural part of software engineering or user interface projects. Some students prefer a more traditional lecture-oriented course with short textbook-based homework problems, but for each of these students, there are others who report that team-oriented projects have changed their lives and that it was the most influential experience in their education. The letters of thanks I get from students, even ten years later, are especially gratifying.

Let's take a hypothetical case of a high school teacher named Andrea and her science students Dona, Raphael, and Michael. They want to enter an online science festival by doing a project on molecular genetics of meningitis, since Dona's father recently recovered from a mild case that he had contracted after a visit to Africa. They enter the terms "meningitis" and "genetics" into LEON's science festival tool, which provides basic science information from encyclopedias and links to authoritative Web sites at research centers such as the U.S. National Institutes of Health, the British Sanger Centre's Wellcome Trust Genome Campus, and the German Max-Planck Institute for Molecular Genetics.

They also get links to seven previous science festival projects that have covered this topic and two groups of students in California and France that are working on related topics. After learning the basics about the bacterial nature of meningitis and the often deadly inflammations it brings to the brain and spinal cord, Raphael dives deeper by following links about the *Neisseria meningitidis* bacteria while Michael talks to his family doctor for basics about meningitis treatments.

Dona finds that the genome of this bacterium was recently sequenced and that the full list of the 2,184,406 base pairs (A, G, T, C nucleotides) has been published online. She excitedly reports this to Raphael and Michael, who find that none of the previous or current science festival projects have studied the genetic sequence of meningitis. They examine LEON's templates for genetics projects that include controlled experiments, mathematical analyses, and structural modeling. Just reading about research methods is informative, and after a week of e-mail exchanges and discussions with friends, they use LEON to develop their proposal to their teacher, Andrea. Andrea likes the basic premise of doing a statistical analysis of genetic patterns in strains of *Neisseria meningitidis,* but she decides to ask the students to get a knowledgeable adviser to help guide them. Raphael uses LEON to identify

appropriate researchers and educators and sends them an e-mail request-
ing consultation. They all reply politely that they are too busy, except
for a retired geneticist in Brazil, who responds positively because his
daughter had died of meningitis. Raphael uses LEON to help verify the
retired geneticist's story and legitimacy, thereby protecting the students.

With the consultant engaged, Andrea approves the project plan but
puts in additional reporting milestones to help organize the work. The
students collaborate among themselves electronically, sometimes late
into the night, with LEON's instant message exchanges as they try a
variety of pattern-matching programs. The genetic structures they find
all match known genes, but Dona notices an interesting combination of
genes that is associated with the gene patterns of her father's youthful
illnesses. His doctor had commented about a family history of childhood
infections. She makes a conjecture in LEON's online notebook: "Could
my father's childhood infections have produced protection against the
more serious forms of meningitis?" Dona doesn't know enough to answer
this question, but she passes it on with her data to the Brazilian consultant, who gets excited about this possibility and reports back to the
group with supportive evidence.

Their science festival project goes through many more stages before
the genetic analyses are ready to be presented online along with the
detailed report. The judges use LEON to review the twenty-seven projects from their high school, checking the logs of activity to ensure that
science festival rules have been followed. Dona, Raphael, and Michael
win a silver medal. They use LEON to send an announcement to selected
researchers, which produces interest from a British geneticist who seeks
funding to follow up on their insight. Their project becomes the foundation for a dozen similar student projects next year. Dona gets an invitation to do an internship next summer at Cambridge University.

THE SKEPTIC'S CORNER

Although educators often portray themselves as open to new ideas,
changing teaching philosophies is controversial. Even though collect-
relate-create-donate is an adaptable teaching and learning philosophy, it
does require unfamiliar roles for teachers and students. Many teachers

and students find it hard to shift from standard lecture formats to the four activities of the new education:

COLLECT   Gather information and acquire resources
RELATE   Work in collaborative teams
CREATE   Develop ambitious projects
DONATE   Produce results that are meaningful to someone outside the classroom

I believe that this philosophy can be interpreted to suit local needs and refined to support collaboration among students. They can learn to communicate effectively with peers and mentors in support of ambitious outcomes that come with service-oriented authentic projects. Learning can become a joyous joint venture/adventure that prepares students for effective participation in communities, for successful employment, and for personal fulfillment.

Still, there are many reasons for resistance to switching to a new philosophy. Change is always difficult, but teachers have good reasons for concern about going too far with collaborative methods. Teamwork can be problematic and is more difficult to manage than individual homework. Measuring learning success for individual students is more complex with authentic team projects because of the inevitable differences in effort among team members and the subjective criteria of quality in creative work. The old-fashioned standardized tests seem safer because of their easy-to-grade objective answers, but they don't measure many important aspects of learning. Explicitly stated criteria for assessment of projects do help, and a blend of projects plus standardized tests is a practical resolution.

A more fundamental concern is that creativity is not universally valued. Many cultures and communities prefer training students to accept existing structures rather than training them to form new ones; they prefer memorization and copying to research and creative writing. These conflicts are likely to remain controversial.

Architectural drawing, from license-free "Leonardo da Vinci: Selected Works," Planet Art.

7

> THE NEW COMMERCE: E-BUSINESS

We must learn to balance the material wonders of technology with the spiritual demands of our human nature.
—John Naisbitt, *Megatrends* (1982), 40

WHY SHOULDN'T YOU GET THE DEAL YOU WANT?

The old business was about making a profit; the new business is about making a profit. This piece of harsh reality suggests that nothing has changed, but of course everyone knows that a lot has changed.

These days, airlines, bookstores, and bed-and-breakfast operators need a Web presence. As a merchant, you can gain a competitive advantage with a strong Web presence, but there are substantial costs to getting started and maintaining the Web site. As a customer, you benefit because you can reserve flights, buy books, or find a place to stay at any time of day and get detailed, up-to-the-minute information. But many people lament the loss of personal contact with storekeepers who offer to deliver heavy packages and bank tellers who remember that you like your cash withdrawals in $20 bills placed in envelopes. These regular contacts help build the social capital that gives many people a sense of community and safety. The storekeeper who accepts returned merchandise beyond the time limit and the bank teller who doesn't need to see your identification card are important figures for elders or new immigrants. Sometimes, trusted relationships are more important than efficient transactions.

But new technology advocates look at the problems they see with service personnel who are unavailable and unpleasant. These technology advocates promote e-business solutions to merchants and argue that Web services will be better, faster, and cheaper. At the same time, these technology advocates tell consumers that they can easily compare prices, ask questions of fellow consumers, or complain online. The first wave of great promises has passed, and now, while the next wave is forming, is a good time for responsible merchants and consumer activists to formulate realistic expectations while trying innovative approaches. Can the Web pay off for a wide range of merchants, or will e-business be the province of corporate procurement officers? Will consumers be protected from unscrupulous merchants whose elegant graphics and well-worded promises are merely a cover for the latest scam? Can government regulation and taxation be avoided?

Since good decisions are based on in-depth understanding, users and designers need to fathom how the experiences of merchants and consumers are changing. Some changes are apparent. For merchants, supply chain and customer relationship management are much stronger concerns. Merchants can keep lower inventories, relying on next-day delivery to satisfy the predicted demand. Merchants value customers but carefully measure how much to offer them. For consumers, greater freedom of choice and novel forms of relationship can be advantages, but it takes skill and time to gain the benefits.

Consumers want low prices, answers immediately, and deliveries yesterday. No technology can satisfy every wish, but let's consider the rhetorical question, Why shouldn't you get the deal you want? To those who remain firmly tied to the old computing ways of thinking, this is a preposterous question. They think customers want to pay nothing and get everything.

The new computing includes consideration of relationships, collaborations, and partnerships that encourage finding win-win deals in which both sides benefit. Reasonable consumers know that companies must make profits to survive, and they want companies to stay in business to provide services when things go wrong and upgrades as improvements are made. Resale values for used automobiles from defunct car makers are much lower than from thriving ones. The deal you want as a merchant is one that creates regular customers who return to make other deals and tell friends of their satisfaction. This emerging notion is sometimes idealistic, but the importance of customer relationships is growing.

Restaurateurs know that customers who feel that they got good value are their best advertising. Satisfied diners will return with friends, to enjoy another meal and because they want to see their local restaurant succeed. Can such personal service scenarios transfer to the world of online e-business retailing and wholesaling? The change of attitude to infuse the Web world with the ancient trust in relationships is incomplete but important enough to be the focus of many books, seminars, and guides to the Web world.

Changes in attitude lead to changes in institutions. The neighborhood flea market where you strolled through crowded booths with cluttered tables picking up knickknacks is giving way to eBay. The used book dealer who held copies of Hemingway's books for you is giving way to an e-mail alert from a used book search engine. Informal chats about your preferences with neighborhood butchers or tailors are disappearing. Your preferences are reduced to a few kilobytes in a database devoted to customer relationship management. Can personal treatment survive? Must a mass society be impersonal?

John Naisbitt, author of the best-selling futurist book *Megatrends* (1982), worried that the focus of some computing promoters was undermining human values. He understood that the new computing would have to blend high-tech with "high-touch," a term he used for personal experience and social contact. His vision was that successful e-business providers would be those who attend to human values while keeping low prices. Naisbitt realized that those who create trusting relationships would be the ones making a profit. Recent writers have reiterated the theme and made customer relationship management the focus of Web-based business strategies (Swift 2000; Lee 2000).

Thinking about Leonardo should also encourage us in the direction of a new computing synthesis of high-tech and high-touch. We can imagine him ordering fancy clothes, like his bright pink knee-length tunic, from Macys.com. We can envisage him selecting canvases and a fancy wood case for his writing implements from MisterArt.com. But Leonardo would probably also want a Web where theatrical performances and public encounters would flourish. He would push for e-business with full attention to human needs, where individual styles, personal expression, and creative initiatives were encouraged. Leonardo would fight against one-size-for-all, limited choices, and central control.

Another inspiration from Leonardo would be universal usability, to ensure that all citizens could benefit. Leonardo was comfortable in the court of the Duke of Milan and in the piazzas with the common citizens of Florence. He understood the needs of middle-class people such as his family and went shopping in the streets of Florence. Leonardo would push us to ensure that merchants and customers could also protect their privacy. After all, he was fanatical about his own privacy—indeed, he wrote in a reverse script so it would be difficult for others to read his notes.

Reflecting on Leonardo also encourages us to think about infrastructures that support commerce and social spaces. His urban plans balanced the transportation needs of merchants with the desire for lively places for people to mingle and interact. He inspires us to make theater, music, and singing a familiar part of our marketplaces. Leonardo-like thinking encourages us to promote painting, drawing, and sculpture as integral components of every shopping mall. In short, we have to remember to enrich business with art and embed social experiences in commerce.

This chapter considers how e-business influences the opportunities for merchants and could bring advantages for customers, leaving aside business-to-business relationships that also have an important role. It raises the twin issues of merchant-controlled personalization and customer-controlled customization. Then it focuses on the generation of online trust by suggesting what consumers should look for from Web-based e-business providers.

OPPORTUNITIES FOR MERCHANTS

The seduction of e-business for merchants is the belief that it is easier to start a Web site than build a store. The dream of frictionless economics lives on—the low cost of opening a business on the Web attracts small and large players. The variations on the dream are that Web businesses can open up new product lines effortlessly and expand into new markets gracefully, even international ones, by simply editing a few Web pages in Microsoft's FrontPage or Macromedia's Dreamweaver.

The dream of frictionless economics is just a dream, because there are harsh realities such as the substantial costs and complexities of finding suppliers, shippers, advertisers, and customer service staff. Even the effort of creating an effective Web site is severely underestimated, especially if the catalog must have thousands of products with attractive photos, changing prices, and up-to-date information on availability.

The illusion of ease of starting a Web business is like a child's fantasy of opening up a lemonade stand in front of her house. It takes only a few minutes to make a sign, prepare some pitchers of lemonade, and move the kitchen table outside. She might even sell a few glasses on a sunny day to sympathetic family and neighbors, but that doesn't mean she is on her way to becoming the next Coca-Cola. Getting serious about advertising, especially when there is competition, requires serious work, and it is not easier in the Web world than in the real world.

However, the Web does have some unique aspects that can provide the advantage necessary to make a success story. Web sites are open twenty-four hours a day, seven days a week, so customers might be lured in, won over, and sold goods at any time. Customers place orders against pictures, thereby reducing the need to keep inventory. This virtual store experience also means that merchants get orders and verify payments before the goods are shipped. This means that there is little loss and no delay in collecting payments.

Forming an activities and relationships table may help merchants spot new opportunities (table 7.1). The ultimate market niche is the individual consumer for whom customized offers can be created on the fly. This form of just-in-time advertising can be based on profiles of the customer that cover age, gender, income, education, city of origin, city of residence, and a hundred other variables. It may be created at the moment of purchase of one product, such as Amazon's suggestion that other purchasers of this book have also bought certain other books. Personal sales strategies can also be based on patterns of previous purchases, Web sites visited, or software used on each customer's computer.

Merchants could move down the rows of the activities and relationships table and focus on small clusters of friends and families with favorable characteristics. This targeted advertising is just what clever travel agents do when they pursue customers with high incomes who are frequent international travelers.

MCI Communications capitalized on this approach in its Friends and Family loyalty plan. MCI offered reduced telephone rates to the twenty people a person called most frequently. Other companies targeted special deals for corporations, employees of those corporations, members of frequent flyer plans, and professional societies. Marketing to different channels has become an art. A data-mining program that identifies target audiences is a good starting point, but understanding and addressing their needs is what the new computing is about. How can you save them time or simplify their lives? How can you make them happier or more secure?

An obvious but unique aspect of the Web experience is that information collection is immediate and cheap, meaning that shrewd merchants can spot trends, subtrends, and sub-subtrends. If twelve-year-old girls in Cincinnati start buying turquoise-colored bandannas on Tuesday afternoon, by Tuesday evening banner ads and e-mails to every twelve-year-old girl in Cincinnati could start a craze and create a market moment for the savvy supplier. Promoters of trips to Cancún, advertisers for

A R T	COLLECT Information	RELATE Communication	CREATE Innovation	DONATE Dissemination
Merchants and customers	Compare prices	*Negotiate deals* Negotiate deals		
Family and friends	*Household patterns* Trusted recommen-dations	*Loyalty plans* Form buying groups	*Niche products*	Tell stories
Colleagues and neighbors	*Local buying patterns* Referral sources	*Targeted advertising* Form buying groups	*Novel products*	Testimonials Complaints
Citizens and markets	*Web usage patterns* Consumer information	*Mass marketing*	*Mass products*	Web site for exchanging info on deals

Note: Roles and activities in italics are for merchants; others are for customers.

TABLE 7.1   E-Business in an Activities and Relationships Table

Ford convertibles, and realtors renting ski condos can create and satisfy demand among a thousand specialized market niches. But remember, the downside risks in life also create needs, so burglar alarm installers can use police reports of break-ins to sell in high-crime neighborhoods, and pharmaceutical companies can stimulate sales by reaching asthmatic or arthritic sufferers.

These strategies for personalization are a remarkable change from the older notions of the mass market. In the 1950s and 1960s, network television had three channels that were broadcast to almost everyone. But the Internet enables merchants to reach people with special needs and interests. For companies who send groups to Chicago regularly, an airline can offer a discount rate for an advance purchase of the next fifty trips. For neighborhoods that have a high volume of purchases of Florida fruits, enlisting one person as the drop-off point can reduce shipping costs that lower the price for consumers while increasing income for merchants.

Of course, there are downsides to the online world and the accessibility of markets—scam and spam. Unscrupulous merchants can deceive customers and never deliver or send inferior quality goods, and then disappear without a trace, only to reemerge with a new Web site and a new brand. Consumers must be cautious about online scams by looking for the indicators that build trust (see discussion of trust later in this chapter).

Spam is a pejorative term that has emerged to describe the piles of unwanted sales messages that most people get, offering everything from laser toner cartridges to Mother's Day flower gifts. Such messages are annoying because they intrude on your work and disrupt your attention. You are right to become angry that your time and energy are being wasted. Finding ways to reach only appropriate customers should be a high priority for merchants, as should allowing users to easily remove themselves from mailing lists.

ADVANTAGES FOR CUSTOMERS

From the customers' point of view there are advantages from the new computing approach to e-business and Web shopping. They, too, can apply an activities and relationships table (table 7.1) to reveal what is possible. An individual consumer can more easily collect information to understand competing products and then compare prices, shipping charges, availability, financing, and guarantees. The communication opportunities mean that you can discuss the quality of service with other customers and join chat rooms to pose probing questions. You could even contact your preferred merchants and create your own personalized deals. If you are not happy with the deals from one company, it is relatively easy to explore other companies and see what their prices are, or go to consumer-oriented product comparison sites like CNET to see a wider range of options.[1] In the new computing, customers should be able to get the e-business deal they want.

At CNET customers can get immediate comparisons on prices, delivery times, and guarantees from five to fifteen suppliers of each computer or electronics product. This gives them enormous leverage when they buy on the Web or walk down to their local store with a printout in hand. This approach probably has its biggest effect on big-ticket items such as automobiles or mortgage loans. The customer is more informed about the possibilities, and confidence in bargaining is heightened. You may often be a rational buyer, but you should also be aware of your emotional state as a negotiator. Knowing about different sellers can make you a more effective buyer.

Getting the deal that you want was the provocative promise of e-business. Cynics would say that the deal any customer wants is to get any product for a dollar, delivered today, with a lifetime replacement guarantee. Reality sets in because a company that offered such a deal wouldn't be around to support the guarantee. A more realistic promise would be to get as good a deal as anyone else got while respecting the needs of your supplier. This is a more complex statement suggesting the collaborative nature of the new computing, but it hints at ways in which the Internet can help. You can discuss deals with other buyers, find out what kind of deals have been made, and understand what the supplier can offer. Then at least you can avoid being ripped off and log off knowing that you made a reasonable and fair deal. You can also collaborate with the merchant by negotiating for a lower price for multiple purchases or if you are willing to take delivery a month from now.

The new computing lets a customer create even more than deals. You can create the products you want. You can custom-order just what you want. You can choose the features and colors for your new car from Detroit or get silk weavers from Thailand to make shirts with your designs. One weaver even offers to let you choose the dye colors, weaving pattern, and details of shirt style. You can create your own store online and sell your creations, while depending on suppliers to produce and deliver the product. However, you will still need to deal with the non-trivial problem of formulating a Web marketing strategy to disseminate and promote your design.

The opportunities to create and sell travel packages, educational materials, or movie reviews have been noticed by many entrepreneurs, who have started their own businesses and become merchants in addition to being customers. But since there is usually competition, it may not be as easy as you thought to make a livelihood from selling your innovations. The Web is the natural place for information-oriented businesses, so offering expert advice, writing electronic newsletters, and starting magazines are natural business ventures. Finding the right niche makes the difference between failure and success.

As a customer, your choices grow. Tracking down the local services that have low shipping charges or offer an opportunity for a face-to-face meeting is the next stage in search design. One service, i411, has a novel Yellow Pages directory that helps you find a dry cleaner or a doctor in your neighborhood or anywhere across the United States.[2]

Selling relationship services is a thriving industry, with referral services of many kinds, often facilitated with personal e-mail to reach the right people. Selling innovation and dissemination services to individuals, small groups, and large companies continues to grow. It even becomes part of national markets as computer and car manufacturers let you design your own products online.

Early Web analysts talked of disintermediation, the elimination of the middle level of sales people because customers could negotiate directly with suppliers. The efficiencies of disintermediation were substantial for some products, especially highly standardized ones, but new opportunities for reintermediation have emerged. The clever traveler can become an entrepreneur by creating a packaged guided tour to trek through Nepal. Combining flights, hotels, guides, restaurants, and even clothing can make an exotic journey seem easy. Then the Web helps to sell the package to youthful adventurers anywhere in the world or to college students in one city who can save money by flying on the same charter flight.

Many variations on this theme seem possible in the form of neighborhood buying services tailored to a community's needs. Advertising and marketing are still needed, and the social process of building trust with suppliers and customers may not be easier on the Web. Social processes are still an important ingredient because trust is slow to build and easy to break. Some suppliers and merchants just don't live up to their promises.

If you are not happy with a purchase, you can complain at the Web versions of traditional groups like Better Business Bureau Online ("Promoting trust and confidence on the Internet").[3] Another option is to go to new online complaint Web sites:[4]

> eComplaints.com is your chance to fight back. It's your only chance to be heard by the company at fault and, more importantly, by your fellow consumers. After all, you don't want any one else to go through what you did. So go ahead: issue a complaint. We'll publish the complaint on the Web and send it to the company. Not only is this more likely to get you a reply, but we'll use the information you give us to help the company improve its service.

As in many arenas, the Internet amplifies the power of individuals to get what they want from big companies or organizations, whereas in the past there were fewer options for unhappy customers. Of course, it remains to be seen how effective such Web interventions will be, as there is little incentive for unscrupulous merchants to bother responding to complaints, especially if they are located in a distant country. The ultimate consumer complaint, the class action suit, also becomes facilitated by network access. Lawyers can more easily identify and reach the participants in a class action suit.

E-business based on the new computing enables you to create your own deal, bargain over prices, or start a bidding war among suppliers. The social space of the Web offers new possibilities. One approach is reverse advertising, which allows you as a consumer to let merchants know you are interested in making a purchase. This is the strategy of Priceline, which enables you to "name your own price" and then let competing airlines decide if they are ready to offer a seat at that price for that date.[5] Priceline requires your payment before the exact flight is offered, but for those with flexible schedules it can offer substantial savings.

This reversal of role and negotiated price is built on the activities of relationship and creativity. Instead of collecting information on what is available, customers communicate with suppliers and create a new deal. Of course, enterprising customers could bargain for a block of tickets and resell them to others at a profit. Sometimes this might be a vacation excursion for neighbors or colleagues. The ease of finding new markets stimulates creative possibilities.

An even more radical approach is group shopping, which develops the idea of disseminating a good deal among friends or neighbors. The co-op, or cooperative buyers network, tried to make this work in the non-Web world, but the idea becomes more viable on the Web because communication is easy. CNET's Ultimate Guide to Group Shopping describes it this way:

> There's strength in numbers on the Internet, especially when it comes to everyone's favorite cyberactivity: shopping. Learn how to join forces with other shoppers at group buying Web sites— online retailers where you pool your resources with other people's in order to buy items in bulk at cheaper prices—and you can receive discounts usually reserved for volume purchases in the off- line world.

Buying cooperatives have always been around, but the Web makes it eas- ier for like-minded buyers to join forces to get what they want at a good price.

PERSONALIZATION AND CUSTOMIZATION

One of the unique features of the Web is its ability to generate a tor- rent of data about every user click. This Web log data, also called clickstream data, can leave a voluminous record of your visits to a

Web site. These data present serious privacy concerns but enable merchants to watch where you are going and then tailor their Web presentations. This is a modern version of what good store managers do. They place staples like milk at the back of the store so that customers have to walk through the whole store, picking up some extras on the way. Similarly, special offers can be stacked high near the checkout section so that everyone sees them. Other opportunities are for similar goods, like low-salt products, to be grouped together. The goal of the merchant is to increase sales, particularly of high-profit items. This can work well for customers, too, who may be happy to learn about special offers and to see novel products on the same shelf with their favorites.

Web merchants have one important advantage over the managers of physical stores. On the Web, customers can be given their own layout based on (1) who they are, (2) their history of shopping patterns, and (3) their selections today.

This Web advantage has given rise to a whole industry of companies providing analyses of user patterns so that merchants can "personalize" Web sites for individual users, recognizable groups, and larger organizations. A competing philosophy is to allow users to customize their own Web sites and choose what they want. Personalization advocates argue that users are too lazy or lacking in knowledge to be able to customize, and that merchant control will maximize benefits. Customization advocates believe that personalization programs often go astray, thereby missing chances for sales, while confusing customers by changing Web sites.

An industry Web site defines personalization strategies: "Marketers using Web and e-mail personalization technologies have the ability to tailor each page seen by prospects and customers. In doing so, marketers can achieve the benefits of using individual salespeople with the cost of traditional mass marketing."[6]

Scenarios based on customer demographics are easy to imagine. Wealthy shoppers can be shown high-value products with established premium internationally known brand names, while poorer customers can be given discounts on lower-quality local brands. A $60 Waterford crystal goblet might appeal to the wealthy shopper, whereas a $1 plastic glass might be just fine a poorer customer. Mature visitors to an online music site might be offered Beatles or classical samples, and trendy teens might get Britney Spears. The urban apartment-dweller will get offers for closet organizers, whereas the suburban home-owner will get an invitation to try a lawn mower.

These scenarios depend on merchants' having information about the customer demographics. This information can come by having customers register and answer questions about income and where they live, or by statistical assumptions based on their zip code. This approach is risky because the merchant may make incorrect assumptions about customers. Some customers living in wealthy neighborhoods are penny-pinchers, and some poorer people like to save up and buy Ralph Lauren Polo shirts. I believe that the more appealing approach, customization, is simply to let users choose what kind of products they want. Giving users control is often a winning strategy for everyone.

A second strategy is to base personalization on individual customer histories of purchases. If a customer buys three books on taking care of new babies, then in a few months suggestions for books about one-year-olds might elicit a purchase. If a customer buys shares of IBM, Intel, and Microsoft, then other computing stocks could be featured, but a mention of diversification to Coca-Cola or Disney stocks might also be appropriate.

A third strategy is to base personalization on selections made during the current session. Old ideas, such as suggesting jelly to peanut butter purchasers, have been expanded with many sophisticated data-mining programs that recognize patterns of purchases in shoppers' baskets.

Amazon has explored many of these possibilities and found a successful formula. I am greeted on one Web page by "Hello, Ben Shneiderman. We have recommendations for you" (figure 7.1) and on another by "Hello, Ben Shneiderman. Explore What's New for You Today." I clicked and found a set of recommendations tied to my recent purchases of CDs and books, plus a few special offers. Then there was the "Movers & Shakers: The Biggest 24-hour Gainers in Books," which highlighted popular books and invited me to join the action. I also had lots of opportunity to specify what I wanted to be informed about by e-mail. Then, when I found a book by Don Norman that I was interested in, it was followed by this teaser: "Customers who bought titles by Don Norman also bought titles by these authors."

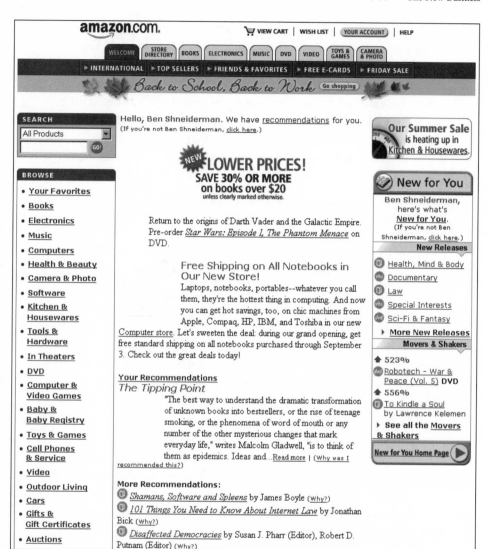

7.1   Amazon.com® Web page, <http://www.amazon.com/>. © 2001 Amazon.com, Inc. All rights reserved.

Some users are delighted by the clever suggestions made by personalization programs, but others are annoyed or frightened. The happy customers are ready to become spontaneous buyers and don't mind if some of the suggestions are not appropriate. The unhappy customers are annoyed by the distraction, the pushy approach to selling, and the invasion of their privacy. It can be frightening to know that your record of health, travel, investment, or erotic book purchases might be available to insurance companies, business competitors, or family members.

Once merchants choose a strategy for personalization, there are complex questions about how to apply it. Little is known about customer reactions to seeing a familiar Web site that they come to know during their repeated visits versus the novelty of an ever-changing arrangement with new offers each time. Most merchants chose a stable layout with some sections devoted to special offers for all customers and others devoted to personalized content for each customer. Merchants who offer users the chance to customize their home page transfer the decisions to the users. For many users, this is the right approach, because control over their environment is a strong desire. When users recognize that the Web site managers are changing the layout in unpredictable ways, they worry about what might happen next. Of course, communicating changes by marking sections with time-sensitive headers, such as "Today's Sales," "New Products," or "Offers for You" help make the design comprehensible.

A key design issue for product catalog Web sites is how close to the home page is each product. Placing a product on the home page of a Web site with 30,000 products is obviously a strong endorsement. Empirical studies show that the more clicks it takes to find a product, the less likely it will be found. Requiring fewer clicks is almost always better. More categories on each page helps users find their way in fewer steps with fewer mistakes. Even though this makes for "busier" pages, which can be troubling to first-time visitors, the benefits of fewer steps to find products are clear.

Highlighting products is also an art form. Featured products can be given prominence by placement (higher up is usually better), size (bigger is better), and emphasis (gold borders or bright backgrounds can help). These and other strategies for helping customers find products they want and steering customers to featured products help make the competitive edge in e-business.

Social traditions are designed to elicit trust during uncertain encounters. Even before Leonardo's time, handshaking demonstrated the absence of weapons. Clinking of glasses evolved from pouring wine back and forth to prove that it was not poisoned. Now, new social traditions are needed to enhance cooperative behaviors in electronic environments that support e-business, e-commerce, e-services, and online communities (Preece 2000).

When you shop online, you can't savor a cup of tea with an electronic rug merchant, so designers must develop rapid strategies for facilitating e-business and auctions. Since you can't make eye contact and judge intonations with an online lawyer or physician, designers must create new social norms for professional services. Since you can't stroll through online communities encountering neighbors with children in strollers, designers must facilitate the trust that enables collective action.

The political scientist Eric Uslaner (2001) calls trust "the chicken soup of the social sciences. It brings us all sorts of good things—from a willingness to get involved in our communities to higher rates of economic growth . . . to making daily life more pleasant. Yet, like chicken soup, it appears to work somewhat mysteriously." He tries to sort out the mystery by distinguishing between moral trust, which is the durable optimistic view that strangers are well-intentioned, and strategic trust, which is the willingness of two people to participate in a specific exchange.

Trust is the facilitator of cooperative behavior. It is a complex concept that has already generated dozens of doctoral dissertations, not only in sociology and political science but now in information systems research. There are enough dimensions to trust and its failures to keep scholars and philosophers busy for some time, but e-business, e-commerce, e-services, and online community designers need a guide to practical action.

The designer's goal is to quickly engage the users, establish strategic trust, and preserve it under challenging situations. But for many users, strategic trust is difficult to generate, easily shaken, and once shaken extremely difficult to rebuild. Strategic trust is fragile.

The large literature on trust offers multiple perspectives. Fukuyama's politically oriented book *Trust* (1995) defines trust as "the expectation that arises within a community of regular, honest, and cooperative behavior, based on commonly shared norms, on the part of the members of that community." This compact definition embodies key concepts: trust is about the future, and it is concerned with cooperative behavior.

In shifting to electronic environments, the Stanford University researchers Fogg and Tseng (1999) focus on trust among individuals mediated by technology. They state, "Trust indicates a positive belief about the perceived reliability of, dependability of, and confidence in a person, object, or process." To separate out the trust for a person or organization from expectations about an object or process, I use the term *rely on* or *depend on* for the positive expectations about a process or objects such as computers, networks, and software.

Computer scientists have concentrated on building reliable equipment; now e-business and e-service providers are trying to win your trust. They want you to get past your hesitation and type in your credit card numbers.

Focusing on the strategic trust that a person has for another person or an organization highlights the distinct nature of human–human relationships. Corporations are legal entities, and there is a long history of resolving problems between individuals and such organizations (which ultimately consist of other people). This leads to my definition: *Trust* is the positive expectation a person has for another person or organization that is based on past performance and truthful guarantees.

Trust is about expectations of the future. Trust accrues to individuals and organizations because of their previous good works and clear promises. It implies responsibility for behavior and willingness to make good for failures. Trust is stronger than reliance because of the responsibility and guarantee. If users rely on a computer and it fails, they may get frustrated or vent their anger by smashing a keyboard—there is no relationship of trust with a computer. If users depend on a network and it breaks, they cannot get compensation from the network. However, they can seek compensation from people or organizations that they trusted to supply a correctly functioning computer or communication service. Understanding the explicit and contractlike nature of trust between people and organizations leads to clearer rules for you as a user of e-business, e-services, online communities, and other Web sites. Clearer understanding will make you aware of what to look for in such Web sites.

Is There a History That Inspires Trust?

You should participate in Web transactions and relationships only if you receive strong assurances that you are engaging in a positive relationship. Seek reliable reports about past performance and clear statements of future guarantees. Seeing a familiar brand and logo generates trust because established companies are usually more respectable and trustworthy. But that is only one starting point because deception is also part of the Web world. As a diligent consumer you should take the following steps.

Investigate Patterns of Past Performance

Airlines report on-time percentages for flights, and realtors advertise how many homes they have sold. Such periodic self-reports of performance may attract users and inspire trust in future performance, as does information about the organization, its management, employees, and history. You should look for clear reports on past performance and investigate records of consumer complaints.

Check References from Past and Current Users

Most people choose doctors by asking for references from friends, but Web-based medical services are likely to be chosen by reading online comments from patients. One of the reasons for eBay's success with online auctions has been its thoughtfully designed reputation manager (Feedback Forum), which enables purchasers to record extensive comments on sellers.[7] Here are some of the 988 positive comments from 978 unique buyers that I found about a seller of cameras:

> *Praise:* Everything works. Quick shipment. Thank you. Peace. A+
> seller
>
> *Praise:* Item exactly as described, fast shipping, smooth transaction.
> A+++++++++++
>
> *Praise:* Fast shipping, camera is just as promised, I'm very satisfied.
> A++++
>
> *Praise:* Excellent service and fast shipping.

There were only three neutral and four negative comments—a remarkable record that inspires trust. However, you will find that any
sellers have excellent reputations because they work on getting buyers
to submit positive comments and because unhappy buyers appear to
be reluctant to submit negative comments. Of course, reputations
can be manipulated, and sellers with poor reputations can reappear
with new identities, so completely reliable reporting is still a distant
goal.

Get Certifications from Third Parties

Lawyers, doctors, and other professionals are certified by appropriate review boards, which may soon certify their online services. Seals of approval from consumer and professional groups (e.g., American medical
or bar associations) help establish trust by third-party reports. Look for
logos from TRUSTe and BBBOnline, which are third-party services
that review online privacy policies.[8] But beware, because review of the
policy statement does not guarantee that the company follows it. Check
for positive certifications, and then visit the complaint Web sites for
negatives.

Are Policies for Privacy and Security Easy to Find, Read, and Enforce?

Privacy policies have become widespread, but some are so hard to find
and read that they undermine trust. Good policies are enforceable and
verifiable. Expectations are rising rapidly as consumers become informed.
You should gravitate to Web sites with well-designed policy statements
accompanied by reports on effective enforcement. If your privacy is a

concern, become familiar with the issues at Web sites such as the Electronic Privacy Information Center.[9]

Is Responsibility Clear?

As you seek a product or establish a commercial relationship, you should expect clear statements of responsibilities and obligations. A well-designed Web site will have orderly structure with convenient navigation, meaningful descriptions of products, and comprehensible processes for transactions. Good design can inspire trust. Simple statements of who does what by when should inspire your confidence. For example, you might prefer a seller who promises free shipping or shipping orders within twenty-four hours of receipt of payment. An auction service that has dispute resolution policies and provides mediation services will reduce the number of unhappy users. Restaurateurs who offer free desserts when dinners are delayed know that prompt apologies and sincere efforts to repair problems plus compensation for failures can win customers for life. Shallow commitments and broken promises should be a warning to stay away.

Seek to Clarify Each Participant's Responsibilities

As with any contract or agreement, full disclosure in comprehensible and compact terms helps build confidence and trust. When terms for transactions, such as price, delivery time and cost, taxes, fees, and return policies are spelled out, you know what to expect and are not shaken by unpleasant surprises. Similarly, policies for communities, such as how long logs are kept, who has access to archives, and what limitations exist for threats or libel, should make you feel safe to have a more open discussion.

Expect Clear Guarantees with Compensation

Since all Web providers are relative newcomers, they must overcome resistance to change and specific fears about credit card abuse, privacy invasion, security risks, and interface failures. Guaranteed protection from credit card fraud is a necessary but not sufficient starting point. Compensation for delayed delivery is relatively easy to specify, but reputation records, authentication, and escrow, admirable parts of eBay's Safe Harbor should be included in Web sites you use. These are the hallmarks of trustworthy merchants.

Look for Dispute Resolution and Mediation Services

Inevitably you will run into a product or service that disappoints you. A crushed delivery box, a delayed medical lab report, or a breach in privacy can all make for unhappy experiences, but the real test is how the merchant deals with these problems. Customer service managers earn their salaries by handling unhappy users with a smile, but you should look for sincere efforts to satisfy your needs and win your loyalty. Well-organized customer service should be standard, and third-party facilitators and mediators are an appealing alternative. Here's the pitch from a online dispute resolution service:[10]

> Resolve disputes over online transactions with SquareTrade's simple, fast, and fair Online Dispute Resolution (ODR) Service. Whether you're a buyer or seller, our service can help you settle a dispute for a fraction of the time and cost involved in traditional legal methods. ODR is completely Web-based and capable of handling disputes between parties based in different states or countries.

These guidelines are merely a starting point for you as e-business participants. A cautious and informed consumer is usually a happy consumer.

THE SKEPTIC'S CORNER

Can the new computing influence e-business enough to make it a success story? There seems little danger that e-business will fade away, but there are risks that it will be undermined by many illegitimate merchants whose deceptive practices scare away the good customers. Gresham's law says that bad money drives out good money. Similarly, bad e-business sites may drive out good ones. Users might become so disillusioned with Web scoundrels, offshore hackers, and network nuisances that even reputable providers will have trouble finding customers.

Another worrisome scenario is that only large corporate sites will be able to become effective e-business providers, driving out the small newcomers or buying them out to reduce competition. A final fear is that governments will intervene to regulate or tax e-business, thereby limiting competition and innovation.

Countering these gloomy scenarios is possible, especially if users and developers attend to human needs. Broadband networks and vast Web server "farms" are the necessary products of the old computing, but it will be the sensitivity to the new computing that leads to happy outcomes for merchants and customers.

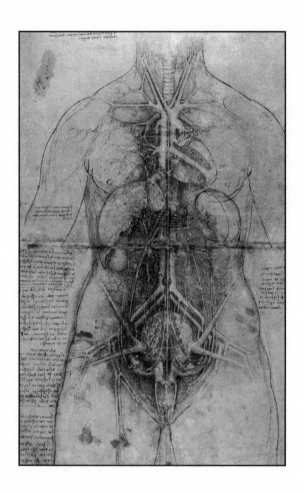

Drawing of a Woman's Torso. From license-free "Leonardo da Vinci: Selected Works," Planet Art.

8

> THE NEW MEDICINE: E-HEALTHCARE

Although Leonardo initiated his anatomical studies in order to enhance his art, they in time became an enthusiasm unto themselves, and finally one of the major endeavors on which his genius was focused.
—Sherwin B. Nuland, *Leonardo da Vinci* (2000), 10

WHY SHOULD YOU EVER BE SICK?

This provocative question is meant to stimulate fresh thinking. Disease has always been a tragic part of human existence, but more diseases are under control now. In the nineteenth century it would have been unthinkable that polio, tuberculosis, or malaria could be prevented, but we now expect this. How many years till researchers uncover the genetic processes that cause AIDS and certain cancers? We may never prevent every disease, but thinking about how to reduce the prevalence of sickness could stimulate novel thinking that breaks from the past.

Medical research is central to reducing illness, but information and computing technologies play key supporting roles. Preventing accidental injuries, common colds, and food poisoning are important goals to which information and computing technologies can contribute. One of the clear signs of human progress is the increasing life span and the relative freedom from tragic painful diseases for people living in developed nations. Extending these gains worldwide remains a continuing challenge.

The first step is to promote dramatically improved medical record keeping. It seems tragic and almost immoral that we are a half-century into the "computer age" without standard medical records available by network worldwide. By comparison, our charge card and airline reservation records are in far better state than our medical histories. Diners can charge their meals at major restaurants worldwide, and airline reservations are available in seconds at any airport, even crossing systems of competing companies and borders of hostile nations. However, you could be dying in a hospital because the attending physician did not know about your medical history or allergies to drugs.

The goal would be that if you are brought to an emergency room anywhere in the world, within fifteen seconds your patient history is on the screen in the local language. The international availability of medical records, let's call it the World Wide Med, would not only improve patient care and potentially lower costs but would have remarkable benefits to clinical researchers, epidemiologists, demographers, and many others. Of course, remarkable threats to privacy are a vital factor that has slowed development, but other forces are probably as influential. The appropriately conservative medical industry is slow to change, especially when the threat to control and the ownership of your medical records are at stake. By standardizing medical records, your physician or health plan may lose some control over you. If your blood tests, X-rays, sonograms, and full medical history were available anywhere, then the originating physicians might fear losing you as a patient. By opening access to medical records, healthcare providers might fear that mistakes and misdiagnoses could become visible.

However, historical precedents show that if privacy issues can be addressed, open access to standardized records benefits service providers by promoting efficiency and improved service. Eliminating duplicate tests and ineffective therapies would benefit everyone because more patients could be helped, hopefully at lower cost. Building the appropriate hardware, software, and networks for the World Wide Med could expand the size of the computing industry by 20 to 30 percent.

Another goal of the new computing would be to empower patients to learn more about their illnesses and to take greater responsibility for their healthcare treatments. Patients are often steered by their friends and family to find informative Web sites and discussion groups related to their diagnosed condition. Getting medical information is one of the most frequent uses of the World Wide Web, with terrific resources such as the Merck Manual in a Home Edition, with versions for nineteen countries.[1] This extensive hundred-year-old guide to medicine for physicians has been transformed into a readable yet authoritative guide for the general public, with photos, videos, and animations. The widespread use of online communities for patient support groups is also changing the experience of patients and the practice of physicians. Yahoo! Groups has more than 20,000 discussion groups under the heading Health and Wellness.[2] More than half are listed under support groups, and another quarter are for professionals. Some groups have more than 100,000 members, but most have under a thousand members and vary in their activity levels. Of course, just as with previous media, there are misleading Web sites that offer misinformation or promote questionable treatments.

Leonardo lived in a time when medical information was still drawn from ancient sources and in which medical research was scarce. Even the basics of blood circulation or the function of major organs was unknown. His endeavors to understand anatomy and use visualization to support research still provide valuable lessons for medical illustrators and inspirations for scientists. Leonardo participated in as many as thirty autopsies to study the muscles, bones, and circulatory system but felt the need to be secretive about this work because such medical explorations were not widely accepted. His capacity for visual thinking and representation enabled him to draw details that were not observed by anatomists for another two hundred years. Historians wonder how medical progress might have been accelerated if his work had been more widely disseminated (Nuland 2000). Leonardo came close to figuring out the heart-lung circulatory system that William Harvey (1578–1657) is credited with explaining. He was the first to accurately draw an S-shaped spine and to understand its importance. His inquisitive approach anticipates the more active questioning of today's superpatients, who are increasingly likely to challenge their physicians and formulate their own diagnoses.

Leonardo's heritage also inspires a rethinking of the goals of the old computing, such as medical diagnosis programs that perform as well as the best doctor. The replacement notion—the mimicry game—has not been sufficiently fruitful because it sets too low a goal. A grander goal of the new computing is to enable the average physician to perform diagnoses far better than even the best physician.

ENABLING PHYSICIANS

Enabling physicians to perform a thousand times better might be achieved by creating comprehensive clinical databases of patient histories, validated simulation models of disease patterns, and collaborative software that allows easy consultation.

Rapid access to the World Wide Med in emergency rooms, clinics, and physicians' offices would enable healthcare professionals to know your existing conditions and compare current electrocardiograms with a historical record to identify changes. As patients visited specialists, sought second opinions, or moved to other cities, their medical histories would be continuously available in a complete and accurate form. This is in dramatic contrast to current practice, in which patient records typically remain at multiple physicians' offices buried in walls of color-coded paper folders, inaccessible to others and often misfiled. Patients' annoyance grows as their visits to new physicians or specialists begin with filling out a lengthy medical history in yet a new format.

This was my experience in attempting to find pain relief for my eighty-five-year-old mother, suffering from shingles. The New York

University Pain Clinic sent a twenty-page form to fill out, but I could hardly feel I provided an accurate or complete record. Then, on entering Beth Israel Hospital for treatment, there was yet another tedious, time-consuming, and flawed attempt to provide background data. My frustration at trying to fill out the forms was accompanied by my recognition that the results would not be reliable, that the bulging files at each site would be costly to maintain, and that the information would probably never be read by anyone. I thought my time could have been spent in more productive and care-giving ways.

Similarly, when my father visited a new specialist for his deteriorating knee, he was sent for a fresh X-ray and then had to return for a second appointment a few days later. Access to previous X-rays would not only have saved the cost and difficulty of making several visits but also have avoided the delay in diagnosis. Furthermore, the physician would have had a better understanding if she could have seen the full sequence of X-rays over the previous few years.

Other benefits would evolve as standard records become widespread. For example, when making treatment plans, wouldn't physicians benefit by having accurate statistical data about how similar patients fared during the past year? Currently physicians rely on distilled information from clinical trials with small populations. Imagine if physicians sitting at their desks could quickly review the outcomes of a thousand patients with similar conditions and even contact physicians who had high success rates. We can only guess what would happen to medical care if physician performance were made available in the same way as mutual funds, baseball players, or airline on-time performance. Why shouldn't physicians be evaluated like most other workers?

In addition to helping individual physicians develop treatment plans, the World Wide Med would be a remarkable resource for clinical researchers. For fifty years the Framingham, Massachusetts, study of 10,300 New Englanders provided a rich database against which researchers could test hypotheses and hunt for correlations. In addition to external benefits the participants received effective continuing medical care that probably improved their health by promoting good diet, exercise, and preventive medicine. Why not make every citizen around the world a participant in national versions of the Framingham study? Of course, some smaller countries with national medical care, such as Israel and the Netherlands, have made progress toward this goal, and they can offer some guidance for other nations with more fragmented systems of diverse healthcare providers.

In the United States, the dream of accessible medical records is stuck in the early stages by massive debates over standardizing formats and terminology. Competing factions have been slow to agree on whether they are pursuing "computer-based patient records (CPR)" or "electronic medical records (EMR)" and on what terms to use for diseases, treatments, or medications.[3] Without public pressure there are few incentives to accelerate progress. The payoffs for patients in improved medical care are not perceived by decision makers as valuable enough to justify the cost or the risks.

Improved data mining from networked patient information could make possible early detection of epidemics or food poisoning. Dr. John Snow's handmade map of dozens of cholera cases in London in 1854 is a guide to what might happen on a daily basis in the future. His marks of the location of cholera patients on a street map quickly identified a specific well as the source of the contamination. By coordinating data from many physicians the patterns of disease could be detected earlier. Yearly patterns of influenza or other contagious diseases could be recognized rapidly, and inoculations or preventive measures could be instituted.

In one heavily publicized event, two children died and many became seriously ill until it was determined that tainted meat was being sold at a fast-food restaurant. It seems likely that many more food poisoning outbreaks go undetected because there is no effective means for finding emerging patterns of illness. Similarly, bioterror threats, such as the use of anthrax, might be detected earlier if shared databases were in place.

It may be standard business practice to go where the market is, but sometimes visionary thinkers can create a market. Medical professionals are not pushing for the World Wide Med, so some groundwork needs to be done. In fact, there is likely to be resistance to this innovation from doctors, healthcare providers, and industry associations. A likely starting point would be to build test systems in selected communities to work out the many difficulties and refine the designs. After three to five years of testing statewide, national and international systems could evolve over the next fifteen to twenty years. This time frame is necessary for the development of these massive systems, the education and engagement of medical practitioners, the change in practices by medical administrators and insurers, and the revision of expectations from patients. Indeed, the social changes in several professions will be more difficult than the technical implementation issues.

Resistance is likely to come from many directions. Physicians, long accustomed to controlling patient history data, may feel threatened if their data and diagnoses become more widely accessible. It will take strong pressure from patients, administrators, and insurers to change this basic aspect of patient-doctor relationships. But change is happening from many sources, such as the Web-based project of Maryland's Board of Physician Quality Assurance, which provides information about each physician's training, affiliations, and performance.[4] The board also provides complaint forms online. I was pleased to see that there were no complaints filed against my doctor during the past ten years.

Privacy concerns are the central question. Can medical records be made easily accessible when needed and authorized, but protected from unauthorized prying? The failures will receive widespread publicity, and sensationalist journalists and clever hackers will find the weak links. However, it seems that current medical files are poorly protected from privacy invasion, from loss, and from destruction. Many doctors' files are just behind the receptionist's desk, subject to easy access when the receptionist steps out to lunch or vulnerable to after-hours burgling. Computerization of medical records might improve their protectability and maintain a record of who examined each record. The goal for online medical records should be to provide better privacy protection than with paper documents. Maybe Leonardo's secret writing style could inspire a more secure form of privacy for medical records.

Another often-stated objection to computerization of patient records is that it will be difficult to change the way physicians summarize medical interviews using a pen. While paper and pen interfaces are quite effective, they have many serious limitations and inefficiencies. Paper is easy to use, requires no power, and is relatively durable, so computerized patient records will have to be well designed to win physicians over. It is a nontrivial task to replace current practices with a computerized input strategy, but it does seem inevitable that improvements can be made that will benefit the patient and the physician.

Some futurists see voice recognition input as the path to the future, but visual displays on portable tablets with touch-screens are a more likely direction. Visual displays and input by pointing are more rapid and reduce the cognitive burden, as compared to speech input/output. But better than manual data entry, every medical device, from the scales that weigh you to the blood pressure cuff, will become an input device that automatically sends results to your patient record. Physician examinations, interpretations, conjectures, diagnoses, treatment plans, and progress reports would all have to be computerized. Certainly, this is a massive project, but the improvements to care and the cost savings from simplified record keeping should justify the enterprise.

Speedups in data collection and entry seem likely, but the potential for a more thorough in-depth interview could be the driving force. A future physician visit might be facilitated by software tools that verify inputs for accuracy and completeness while prompting physicians about anomalies and reminding them to check all possible diagnoses. These tools must be designed to speed the work of physicians, enabling them to see more patients or to give more attention to patients' questions. Managed care providers will press for greater physician productivity; patient groups must press for more personal attention.

Standard medical records would mean that physicians would not have to repeat questions but could rapidly review patient histories, presented in a standard visual format that identifies key problems. Then, ideally, your physician could probe more deeply about problems and focus on your genuine concerns. Many good doctors take the time to do this, some healthcare providers encourage it, but the pressures of time and profitability are part of the healthcare equation.

The remarkable progress in medical imagery since Wilhelm Conrad Roentgen (1845–1923) discovered X-rays in 1895 will continue. X-rays reveal bone breaks or spinal injuries, while computer-assisted tomography (CAT) scans and sonograms show cancers or fetal anomalies. Sharper images and three-dimensional displays will aid diagnoses and help you understand your problems. Small television cameras used in colonoscopies show polyps in your intestines, but these procedures will become easier as still tinier swallowable TV cameras will broadcast images as they travel through your body. Medical monitoring devices, worn like jewelry bracelets or rings, will continuously record your blood pressure, pulse rate, and temperature, and give early warnings of emerging problems. Other devices will enable you to send data from home to your physicians for a variety of treatments, as many insulin-monitoring patients already do.

Another class of enabling tools for physicians could be advanced simulations of body systems, disease processes, and treatment effects. Measurements of your body functioning, medical images, and genetic profiles could feed into simulations that could show your circulatory problems due to arteriosclerosis or the blockages caused by your sinus head cold. Then different dosage levels of medications could be tested to see their impact and detect potential side effects.

Genetic tests and genetic transfer may lead to breakthrough technologies. Dramatically improved treatments based on deeper understanding of cellular processes could flow from the remarkable efforts of the international Human Genome Project.[5] The ethical questions are troubling, as parents will have to decide what to do if they are likely to produce a child that has genetic deficiencies, and adults will have to decide what to do if they are likely to get breast, ovarian, or prostate cancer. More promising scenarios will emerge as biotechnology moves from detection to prevention and early intervention. Genetic treatments by identifying and replacing faulty genes may eventually send certain cancers into remission or prevent AIDS. These dreams might be realized in the next few decades as information technology supports progress in the biosciences. Major computing research projects, such as IBM's Blue Gene Project, will support biologists who are seeking to understand the three-dimensional structure and function of human proteins.[6] As these puzzles are solved, drugs could be formulated and constructed specifically to fit your genetic profile, although the medical challenges and ethical questions are substantial.

The complexity of modern medicine means that your physician must often collect additional information to make a decision. You may have an unusual genetic history, rare disease, or new medication, so rapid access to electronic sources is needed. The familiar Physicians Desk Reference has been expanded and made available electronically in CD-ROM and Web-based versions that can be continuously updated. But many cases require physicians to consult with, or refer patients to, specialists for novel treatments that require expensive equipment and specialized skills. Improved consultation tools through tele-medicine collaborations are making rapid assessments possible. Imagine that your physician organizes appropriate sections of your online medical history for a presentation to a specialist through the World Wide Med. Then they could have a chat by voice as your physician paged through your medical history and discussed treatment plans. The specialist might want some time to review the materials and consider alternatives. Arranging the technology support will require some innovation to make rapid transfers and smooth discussions possible. But the technical solution will have to be accompanied by deliberation to arrange for payment and liability. A physician (or lawyer or other professional) may discuss cases with colleagues informally while walking down the hallway, in which case no payment or liability is assumed. However, if specialists become involved more seriously, at some point payments and liability arrive hand-in-hand with the need for documentation and billing.

Networked consultations are a middle ground between casual conversations among colleagues and sending patients to meet with a specialist. Such tele-medicine consultations raise administrative and legal questions, especially if the consultant is in a different country where fees may be lower but laws may limit liability. How might the potential benefits to patients and the threats to quality care be reconciled?

EMPOWERING PATIENTS

The future of healthcare will be influenced by how well the technology supports the physicians and other healthcare professionals, but an even stronger factor will be the rise of the superpatient. More and more patients arrive at their physicians' offices with annotated printouts from medical Web sites and e-mailed guidance from friends and family. The subservient patients who accept what their physicians tell them still exist, but the superpatients who are second-guessing their doctors are becoming more common.

A Pew Foundation report in 2000 found that 52 million American adults, or 55 percent of those with Internet access, have used it to find out about a disease or medical condition (Rice and Katz 2001). Of those getting health information online, 48 percent reported that the advice they found improved the way they take care of themselves, and 55 percent said that the Internet improved the way they get healthcare information. Recent reports show still greater activity in the United States and growing numbers internationally (Preece and Krichmar 2002).

Many doctors are happy to see informed patients who take more active roles in their treatment. However, some doctors are not used to the challenging attitude of some patients and are annoyed by their inaccurate, incomplete, or out-of-date information that can undermine the patient-doctor relationship. Studies of medical Web sites reveal the problems of misleading information, but many patients are better educated and informed than in the past.

Sources such as the U.S. National Library of Medicine and the U.S. National Institutes of Health provide excellent information with separate sections written for physicians and for patients.[7] These Web sites cover diseases such as cancer especially well, giving detailed information about the disease, standard treatments, and experimental trials.[8] Corporate medical resources that provide comprehensive consumer-oriented medical information include WebMD and former U.S. Surgeon General Dr. Everett Koop's Web site.[9] You can search on a wide variety of terms from *acne* to *zygote* and find information written in lay terms.

Coverage of more specific diseases comes from professional associations, consumer groups, and individual physicians, related to Parkinson's disease, breast cancer, or dentistry. They provide helpful information to patients by referrals to leading treatment centers and access to recent research results. But they do more than provide information for patients and healthcare providers. Many groups have a political agenda, such as increasing funding for research or changing insurance policies of major healthcare management organizations. E-mail and discussion groups dramatically facilitate organizing thousands of patients and patient rights advocates worldwide. Leaders can conduct discussions to shape policy and then coordinate fund raising, event planning, or letter-writing campaigns. Such community-forming activities can make a difference, as the Parkinson's advocates demonstrated by gaining substantially increased research funding from the U.S. Congress.

However, as every medical advocacy group becomes proficient in the use of e-mail and discussion groups, the advantage of the early adopters decreases. Getting five thousand patients to send e-mail to congressional leaders will be common, and more dramatic signs of support will be needed. Internet skeptics say that the technology changes nothing, and that there are other more important determinants of success, such as famous Hollywood actors who support a disease group or great leaders who promote their causes vigorously. Leaders are important, maybe most important, but the advantage gained by early adopters of appropriate technology can be substantial. Military planners from well before the time of Leonardo have understood this difference, leading to their advocacy of advanced technology. Industrial competitors have also recognized the value of innovation, as demonstrated by patents, new products, and superior infrastructure support. Now citizen groups, especially in the medical arena, are using technology to organize themselves.

Other potent forms of communication are the tens of thousands of health support groups in which millions of patients participate. Patients with rare diseases can discuss their treatment with similar patients around the world by way of the many technologies that support these online communities: chat rooms (figure 8.1), listservs, news groups, and threaded discussion lists. The participants exchange information about physicians, hospitals, treatments, and outcomes. The information interchange is substantial, but the larger payoff seems to be the emotional or empathic support that participants receive from peers.

If you've torn the anterior cruciate ligament (ACL) in your knee while skiing, playing basketball, or just tripping down stairs, then you'll want to visit Bob's ACL Bulletin Board (figure 8.2). His Web site shows somewhat gory pictures of his surgeries and Bob smiling with braces on both knees. He provides basic information and links to online information resources, but the interesting action is on the bulletin board where hundreds of questions and responses are posted every week. This is typical:

> Hey! I was wondering if anyone had a nerve block when they had surgery. Did you have any pain when you had it? Because when I had surgery I had a nerve block and I did have pain. I was just wondering if it was normal or not to have any pain at all since the nerve block is supposed to numb your leg and not have any pain. Very curious, Erin.

Jenny Preece's (2000) analysis of the messages showed that the majority had empathic content, and that purely information transfer or simple narratives were in the minority. Participants described their injury and asked for advice about if and how to have surgery and which physician or hospitals to use. They got lots of advice, and they also got sympathetic

Anne I have been very frustrated myself that I should have problems eating. I wasn't diagnosed until I was 27 with IDDM (I don't make any of my own insulin). I ate very healthy before the diabetes. Suddenly way too much focus of my time is on food.

Andre Hey, I've found that the only way to lose weight is to eat properly and to exercise to get your metabolism moving. I lost plenty of weight two years ago and guess what? I've put it all back on cause I haven't been exercising enough.

Caterina Anne, have you considered an insulin pump?

Anne AND it is very easy (and therefore very tempting) to try to lose weight the wrong way. I'm sure I'm not alone in this.

Michael Anne, do you mean by not taking enough insulin?

Andre You know, after a while being diab you get to know what you can eat and what you can't eat and you can exchange one thing for another . . . In the beginning though it is tough. Not good to occupy your mind on food all the time.

Anne I am on an insulin pump.

Elizabeth I find it frustrating exercising, then having to eat because my sugar drops. It defeats the purpose of exercising, so I don't bother. It's a good excuse, anyway.

Caterina How is it working out? We have just begun the process to get my daughter on one next summer.

Anne Yes, and by eating enormous amounts to induce DKA.

Luke I can see where it would be easy to try and find some good out of a diabetes diagnosis and start to believe that weight loss is perhaps a way to turn something that might be considered bad into something you can capitalize on. Is there an element of that?

Anne Elizabeth, I m glad you brought that up. I had that problem, too.

Andre Anne, if you're on the pump, what is the problem? I mean, you can control your bs's with the pump so much better only taking the insulin you need to take. You're treading on thin ice if you don't take enough insulin, high bs's, possible ketosis and coma

Anne Luke: Very perceptive. I think so. Caterina: Were you talking to me?

Michael I am sure Anne knows that but the head and heart don t always agree

Elizabeth The trouble is, the weight loss isn't permanent, Anne. The minute your sugar is back under control, the weight goes back on.

8.1   Chat transcript (part) from diabetes chat room on WebMD (names have been changed)

Bob's ACL WWWBoard

Message Index

Welcome!

Messages Posted   137 of 3140 Messages Displayed
Within the Last 3 Day(s)   (Reversed Threaded Listing)

- **Constant pain along the medial side of my knee....** -- Susan -- Sunday, 7 October 2001, at 5:06 p.m.
- **Plaster Cast after operation?? Anyone?** (views: 8) -- James -- Sunday, 7 October 2001, at 3:51 p.m.
- **nerve blocks** (views: 10) -- Erin -- Sunday, 7 October 2001, at 3:39 p.m.
  - **yes some discomfort** (views: 1) -- mds -- Sunday, 7 October 2001, at 4:27 p.m.
  - **Re: nerve blocks** (views: 6) -- Debra (Austin's mom) -- Sunday, 7 October 2001, at 3:42 p.m.
    - **Re: nerve blocks** (views: 5) -- Erin -- Sunday, 7 October 2001, at 3:45 p.m.
- **blood clots?** (views: 5) -- Sim -- Sunday, 7 October 2001, at 3:31 p.m.
  - **Tests and treatment** (views: 4) -- Michelle N -- Sunday, 7 October 2001, at 4:00 p.m.
- **shin pain now!** (views: 19) -- dr.anuradha singh -- Sunday, 7 October 2001, at 10:36 a.m.
- **I'm gonna be back in action in a week or two!** (views: 24) -- Joseph -- Sunday, 7 October 2001, at 12:07 a.m.
- **patellar tendon pain** (views: 30) -- oggie -- Saturday, 6 October 2001, at 9:48 p.m.
  - **Re: patellar tendon pain** (views: 1) -- jenna -- Sunday, 7 October 2001, at 4:25 p.m.
- **Knee Braces** (views: 39) -- Amar Dhaliwal -- Saturday, 6 October 2001, at 8:35 p.m.
  - **you got mail...** (views: 26) -- oggie -- Saturday, 6 October 2001, at 10:05 p.m.
- **have a date...** (views: 21) -- stuart2348 -- Saturday, 6 October 2001, at 8:04 p.m.
- **Re: ACL Surgeon UK needed - advice pls?** (views: 12) -- Fiona -- Saturday, 6 October 2001, at 6:46 p.m.
  - **Re: ACL Surgeon UK needed - advice pls?** (views: 5) -- Stephanie -- Sunday, 7 October 2001, at 12:44 p.m.

8.2 Bob's ACL WWWBoard. Bob's Kneeboard is a publishing service of the Factotem Constellation.

responses saying, "Don't worry, I've been through it and you'll do fine." Discussion often went deep into personal fears and extended over weeks and months through surgery and recovery. Those who were early recipients of good advice and warm wishes returned to reciprocate repeatedly over years, inviting newcomers to "let the group know how your surgery turns out—we're cheering for you." AIDS sufferers and leukemia patients have similar Web sites, each tuned to the specific needs of that community, often with separate discussion groups for men or women, old or young, severe or mild cases. Empathy appears to be stronger when participants share common backgrounds or experiences, since they can most easily identify with the challenges that another person faces.

As these groups evolve, they often fragment into smaller, more focused groups where discussions can mature over time, leaders with distinctive personalities can arise, and trusted participants can address new themes. The dynamics of online community evolution are just beginning to be understood, as are the strategies for scaling up from thousands to millions of participants. Preece's framework for understanding online communities is useful to participants in deciding which ones to join and for moderators in deciding how to manage them. She describes these elements:

> *People.* Who are the participants? What is their age, gender, knowledge, location, income, education? How many are long-term or short-term participants? Thriving online communities usually have a well-defined set of participants.

> *Purposes.* What are the purposes of this group? Are the goals clearly defined and shared by the participants? Does the discussion remain focused on the purpose? Successful groups usually have clear purposes.

> *Policies.* Is this a closed or open group? Are discussions archived? Are anonymous postings permitted? Does someone moderate or monitor to keep discussions on topic and prevent excessive hostility or other inappropriate behaviors? What privacy protection is there? Explicit statements that describe management policies, decision-making processes, and dispute resolution procedures are becoming more common.

As online communities grow larger and more important, the usual problems of real communities emerge. Rude or illegal behavior by disruptive individuals needs to be dealt with appropriately if the remaining participants are to feel safe. Newcomers to such online communities should look for policy statements, rules of behavior or etiquette (sometimes called netiquette), or bylaws. Many beginners are reticent about participating, so a comfortable approach may be to join the typically large number of lurkers—those who follow discussions but don't contribute. This is common behavior and is often a good idea to gauge the level of the discussion, style of debate, or receptivity to novices. For most online communities it's fine to be a lurker for a long time, while in some, typically small, groups, active participation is expected.

Once you've determined that the people participating in an online community are appealing to you, that the stated purpose is in harmony with your own, and that the policies make you feel safe, then you can get involved. In small communities with dozens or hundreds of participants the volume of messages or postings might be just a few a day, so it is easy to monitor the discussion. A lively topic can generate dozens of messages a day, but that is still manageable. Some participants get entranced with the discussions and can spend hours a day reading postings or participating in chat discussions. With larger communities there may be thousands of messages a day, so strategies for limiting postings and organizing topics are essential steps.

Medical support communities are especially sympathetic to newcomers and accommodate their needs with pages of Frequently Asked Questions (FAQs) so that the typical inquiries are not repeated to the annoyance of old-timers. There are standard patterns of questions asked by newly injured or recently diagnosed patients. But some newcomers want more than the information in the FAQs, so designated welcomers are ready to offer warmer and more personal greetings with helpful responses, even if the information is already in the FAQs.

Medical discussions may focus on the patient's needs or the caregivers who attend to friends and family with the disease. A growing number of groups focus on political activities to promote increased research funding, change laws, revise government regulations, or change corporate insurance policies. As communities grow still larger, strategies for splitting into smaller discussion groups with summaries will have to be created. Many services fragment groups by ever more specialized topics, geographic location, gender, or age.

A natural expectation is that government and private rating services will emerge for physician performance or hospital satisfaction, such as the Board of Physician Quality Assurance. Collecting comments from many patients could be helpful, but concerns over quality control and legal challenges will have to be overcome for such services to flourish. It may take a while till the lessons from eBay's reputation manager migrate to the medical world.

By now, you may already be thinking in terms of an activities and relationships table for healthcare (table 8.1). Going across the rows will lead you to the idea that a small fraction of devoted patients will write up their experiences in Web pages and books in creative ways to benefit others. Medical books have a large market as a form of self-help to future patients. Such books may cover tragic stories of medical failures or remarkable reports on successful treatments. They often call for changes from medical professionals or organizations that can be constructive and bear fruit. Web-based stories are often a component of medical online communities, providing a therapeutic release to authors and a way of disseminating knowledge to other patients. As the volume of these stories grows, indexing and search methods should enable you to find reports from people of similar age, gender, and treatment. Eventually thematic taxonomies will allow you to find stories that report on the role of sympathetic doctors or dietary changes, even when you don't have a specific keyword to search on.

Another likely direction for medical technology would be to support patient needs for data collection by applying the ideas of InfoDoors and WebBushes. Some patients already collect blood pressure readings daily, monitor their food intake, and run medical tests at home, such as blood sugar levels for diabetics. Facilitating data collection with small information appliances could improve medical care for patients and do much to promote good health. Imagine if your kitchen appliances were equipped with DietTracker sensors, enabling you to monitor your food preparation for cholesterol, salt, sugar, or calories. You could detect foods that were unhealthy for you and discover which vitamins or minerals were missing in your diet. Portable devices would let you spot overly salty or excessively fat restaurant dishes. How about a spoonlike device, HealthSpoon, whose handle has a digital readout of the nutritional values of the food you scoop up. Not everyone wants to know these facts, but an increasing number of people are health-conscious.

ART	COLLECT Information	RELATE Communication	CREATE Innovation	DONATE Dissemination
Physicians and patients	DietTracker HealthSpoon	Personal logs	Personal summaries MedBook	Published memoirs
Family and friends	Family disease histories	*Information consults* Empathic support	Experience reports	Family genetic histories
Colleagues and neighbors	*Specialist resources* Referral sources	*Information consults* Empathic support	*Novel treatments* Experience reports	Lessons learned
Citizens and markets	*Physician information* Patient information	Advocacy groups		*Outcome reports* *Validated treatments*

Note: Roles and activities in italics are for healthcare providers; others are for patients.

TABLE 8.1   Healthcare in an Activities and Relationships Table

Other devices resembling smoke and fire alarms in your home might detect your rising blood pressure or falling iron levels. These devices could alert you to dehydration or cancer precursors in your blood. Advanced devices that you wear could monitor your sleep patterns, exercise efforts, or stress levels. Some devices would log the data for your private MedBook, just as you record your financial activity in a checkbook or family expenditures log. Other devices would be connected to the World Wide Med in order to send the data to your physician or health management organization. Of course security and privacy protection would have to be properly supported.

Empowering patients requires education as well as costly equipment, which is a further reason for concern about disparity in medical treatment. Improving education is a fundamental challenge, and reducing costs through high-volume manufacturing can help, but governmental mechanisms will be necessary to promote equal access to quality medical care. CompuMentor and its TechSoup Web site offer information technology guidance, professional vounteers, and online resources.[10] Since 1987 they have helped more than 23,000 nonprofit groups provide better services in medical, educational, and community projects. The further challenge of bringing improved medical care worldwide should become part of the new computing agenda.[11] The United Nations Information Technology Service, launched in 2000, has similar goals at an international level.[12] Organizations devoted to emergency medical help and improved medical care internationally include the heroic French-founded Médecins sans Frontières (Doctors Without Borders) and Med Help International.[13]

A MEDICAL SCENARIO

Let's envision how your medical visit might unfold in the future. Your physicians would begin with a review of your medical history, examining your recent hospitalization records and conducting genetic lab tests to determine the exact nature of your illness. Then they would access the World Wide Med to collect up-to-date information on success rates of alternative treatments. If your case were something out of the ordinary, they would be able to consult with other physicians rapidly, create novel treatments tuned to your needs, and run simulation studies with different dosage levels. When your treatment was complete, the records would be available to others (while preserving individual privacy), and any novel treatments would be added to the library of therapies available to other physicians.

I hope readers who are healthcare and computing professionals will permit their imaginations to go free for a moment, and come along with this fantasy. We start with a patient, named Dorothy Gale, who has recently returned from her adventure in the Emerald City in the Land of Oz.[14] Unfortunately, she is running a high fever and has a Ruby Red Rash on her feet. Her aunt takes Dorothy to Doctor Louise Pasteur, who examines Dorothy and does a series of blood and genetic tests.

Dr. Pasteur studies Dorothy's medical history LifeLines display (figure 8.3), which shows her previous hospitalizations.[15] Dr. Pasteur examines Dorothy's X-rays, body scans, and electrocardiograms, but these seem normal. The Ruby Red Rash symptom is serious and unusual, so Dr. Pasteur decides to examine a diagnostic reference source, linking to it by way of symptoms and Dorothy's history plus genetic test results.

Many potential diagnoses appear, but the strongest link for Ruby Red Rash is to Munchkin Syndrome, also known as Redmond Rash (figure 8.4). Munchkin Syndrome commonly leads to disorientation and loss of memory. It was discovered in 1982 by Dr. William Gaitze.

To verify this diagnosis, Dr. Pasteur selects the primary symptom—high Toto cell counts—and links back to LifeLines record. Sure enough, Dorothy's LifeLines record indicates that the last blood test showed a high Toto cell count of 81 and another anomaly, a high blood density of 7. This is a somewhat unusual situation, so Dr. Pasteur decides to consult the World Wide Med, containing 631 million patient histories.

She extracts four thousand cases of Munchkin Syndrome and trims the set to 308 cases of those of similar age and genetic background to Dorothy (figure 8.5). Then Dr. Pasteur explores possible treatments from a database of treatment templates. She explores treatments by selecting "chemotherapy", "dialysis", "interferon", and "surgery". Chemotherapy is used only with women, dialysis is the favored treatment, interferon has only moderate impact, and surgery is seen as too risky. She displays the cases with Toto cell initial counts on the x-axis and the amount of reduction on the y-axis. In this Spotfire display, red points are females and blue points are males. Size encodes blood density.[16]

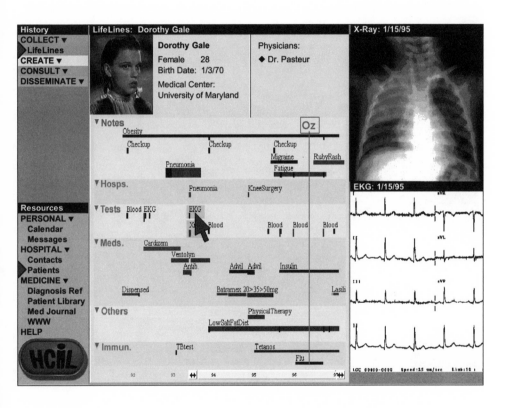

8.3   The fictitious Dorothy Gale's LifeLines medical history showing chest X-ray and electrocardiogram.

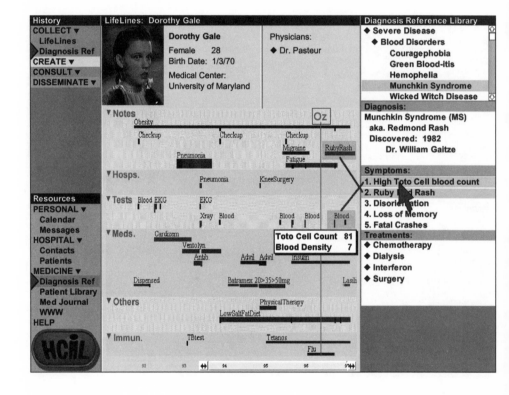

8.4   The fictitious Dorothy Gale's LifeLines medical history showing symptoms of Ruby Red Rash and high blood density, linked to the diagnosis for Munchkin Syndrome.

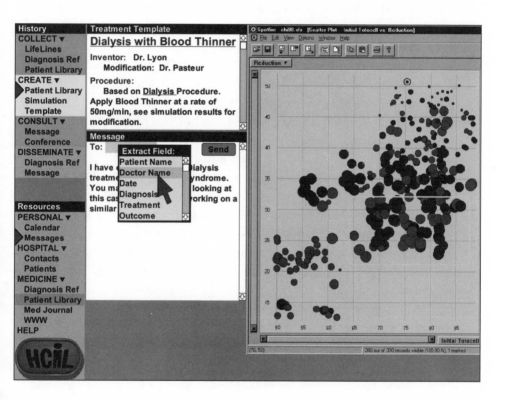

8.5   Send message to Dr. Lyon showing selected cases similar in age and genetic background
to the fictitious Dorothy Gale.

For high Toto cell initial cases, dialysis shows the highest Toto cell reduction for patients like Dorothy. However, Dorothy's family has a common genetic pattern of high blood density that may reduce the effect of the dialysis. Sure enough, a scan through the blood densities shows that those with high blood density do not do as well with dialysis.

But Dr. Pasteur is an expert on blood density and believes that a blood-thinning treatment may make the dialysis more effective. So she runs a simulation that introduces blood-thinning drugs at 10mg/min. The simulation doesn't show a sufficient effect, so Dr. Pasteur decides to select the simulation history and rerun it while changing the blood thinning drugs from 10 to 20 to 30 and up to 100mg/min.

She finds an optimum decline of Toto cells at 50 mg/min, but Dr. Pasteur is worried about the blood-thinning problem and decides to consult with the developer of the dialysis treatment method for Munchkin Syndrome, Dr. Lyon. Dr. Pasteur sends Dorothy's case history as well as the links to Dorothy's record on the World Wide Med and the dialysis simulation results to Dr. Lyon by secure e-mail.

A few hours later Dr. Pasteur has a video consultation with Dr. Lyon, an old friend from medical school days, who is intrigued by the simulation results. He confirms the 50mg/min rate and signs on as a consultant, affirming the treatment plan and taking a share of the income and liability. Dr. Pasteur arranges for Dorothy to get the dialysis, and her Toto cell count drops quickly as the rash fades away. The treatment is a success.

Dr. Pasteur records the positive outcome and then disseminates the results. She adds her revised treatment template to the database and sends an e-mail note to physicians that are treating similar cases to inform them of this case. Then she begins to write up this case for the *New England CyberJournal of Medicine.*

THE SKEPTIC'S CORNER

Medical computing brings out the worriers in droves. Almost any suggestion is met by resistance from some physicians who fear loss of control over their decision making and patient-doctor relationships. Critics of health management organizations fear that changes will only promote a more narrow-minded focus on profitability that ignores patient needs. And even if physicians, patients, administrators, and insurers could agree on how to proceed, privacy advocates see a history of privacy violations with little hope that adequate controls can ever be installed. These are legitimate concerns, but the potential benefits in improved medical care warrant further study and pilot projects to test solutions.

Concerns about medical information quality on the World Wide Web are also appropriate and could result in improvements by more careful monitoring. Techniques for promoting correctness, completeness, and timeliness could be refined. The benefits of empathic support from patient support groups need to be documented, which should also lead to methods for improving their efficacy while reducing the occasions when anger and attacks disrupt discussions. Clear definitions of purpose, people, and policies will help moderators assist newcomers, guide discussions, and limit off-topic excursions or angry flaming behavior.

But even if these steps are taken, there is an enduring challenge in trying to cross the medical digital divides within developed countries and then to developing countries. Justifying the substantial costs for computing technologies is a challenge when there are insufficient medical personnel, and even more difficult when there are none. Even as information and communication technologies continue to drop in price, wise decisions are needed to deploy them so as to maximize the efficacy of the available medical resources. In poorer neighborhoods and developing countries information and communication technologies are more immediately effective in hospitals, clinics, and community centers.

And finally, decision makers should appreciate that computers are merely tools for use by healthcare professionals and patients. They don't substitute for the congenial bedside manner offered by competent healthcare professionals, who make trusting relationships with their patients. Competence and kindness are both valued aspects of the new computing.

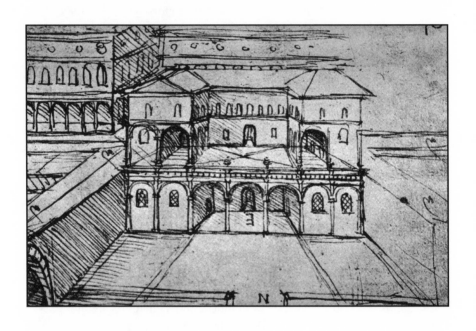

Drawing for the Plan of a Town. From license-free "Leonardo da Vinci: Selected Works,"
Planet Art.

9

> THE NEW POLITICS: E-GOVERNMENT

The most sacred of the duties of a government is to do equal and impartial justice to all its citizens.
—Thomas Jefferson's note in Tracy's *Political Economy* (1816)

WHY SHOULDN'T YOU GET WHAT YOU WANT FROM GOVERNMENT?

Distrust for government has a long and noble tradition, but so does citizen activism and public service to make government more effective. The balance of skepticism and support is always changing, even as citizens' desires to receive government services grows and their willingness to pay taxes shrinks. Since information and computing technologies have restructured the political landscape, finding new ways to reshape government services to better serve citizen needs and understanding how citizens can participate more actively in political processes are once again on the agenda. My descriptions are derived largely from the United States but are meant to apply internationally.

The relationship of information and communication technologies to government and political processes has always been strong. Thomas Jefferson conveyed this linkage in a remarkable statement in which he said that he would prefer having newspapers without government to having government without newspapers. He was making the point that media provide effective means to communicate information and ideas, thus supporting the deliberation that leads to consensus and limiting the excesses of those in power. As much as a free press was seen to be necessary for democracy in the past, a free Internet may be seen as necessary for democracy in the future.

But how free is free? Free speech is a wonderful principle, but it has limits at the borders of slander, incitement to violence, and child pornography. Privacy is an admirable goal, but it can be breached with a search warrant to protect the lives of others or given up casually in exchange for the promise of improved services. New technologies force citizens to redefine the limits of what they find acceptable in terms of free speech, privacy, regulation, and taxation. Many people desire an unregulated Internet until they receive destructive viruses or become the victims of an e-business scam. Most citizens want an untaxed Internet until they realize that their property taxes may have to be raised to compensate for declining sales tax revenues that no longer support desired government services.

While many newspapers are free from government controls, they are businesses with commercial pressures to make a profit. Advertisers who interfere with editorial freedom and the media monopoly owners who restrict criticism may limit the role of newspapers in promoting open discussion. Other limits to the democratic benefits are that most newspapers charge a fee and of course require literate readers. The parallels with Internet news, discussion groups, and online communities are strong, since the same forces prevail. However, the cost of computer resources combined with Internet service fees and the skills needed for computer literacy are a barrier. By contrast, the cost of newspapers and skills to read them are low, thereby raising serious questions about universal access and usability of the Internet.

The promise of the Internet and the World Wide Web to provide improved e-government services is one of their great attractions. Rosy scenarios include citizens searching the U.S. Library of Congress or the British Library to harvest the world's knowledge, making reservations for national park visits, or learning about their retirement rights. Business managers are promised more accurate economic forecasts, farmers are offered satellite images to determine crop status, and urban planners hope to integrate census data with construction permits to identify demographic shifts. These are becoming possible, although currently complex designs limit the number of success stories. Improved data resources and designs for universal usability will eventually enlarge the number of successes and expand the range of possibilities. These improvements will help you get more of what you want from government, but of course there must be limits on what government can give you. And maybe you want less from government—less invasion of privacy, fewer taxes, and less regulation of your business.

Political processes to decide what government will give you are the second major theme in this chapter—how to create open deliberation about what government gives you. This theme deals with how we choose our government leaders through democratic political processes and how we promote our causes with legislators. Information and communication technologies are supposed to facilitate discussion and encourage participation in local, state, and national elections, but the impact has been modest thus far. How might a million people participate in meaningful political discussions that contribute to arriving at mutual understanding and forming a consensus? How might political parties solicit volunteers and contributors? How might citizen groups organize to lobby for change?

L Of course, Leonardo lived in predemocratic times, but he was closely involved with the powerful rulers in Florence and Milan, and served the prominent leaders in Venice, Rome, and France. He needed protection from invaders and support for his projects, for which he offered military innovations, urban designs, and public art. For Leonardo, governments were the source of public works projects to which he contributed in order to earn his income. His efforts at swaying public opinion focused on promoting painting over sculpture, rather than left-wing or right-wing politics. Leonardo's inspiration for the new politics and e-government is tied to his devotion to the public good and to creating satisfying public spaces, for which he made urban designs and provided art. His commitment to appropriate infrastructure as a powerful force that shapes social processes can still inspire us (Lessig 1999).

GETTING WHAT YOU WANT FROM GOVERNMENT

Many citizens expect a lot from government. Citizens expect local governments in cities, counties, and states to take care of local needs for education, transportation, police, water, sewage, and recreation. Local governments have direct immediate impacts on people's lives and elicit strong participation from citizens. Local government services are more likely to be delivered by familiar local officials, who are close to the problems and care about effective solutions—or so the theory goes.

Citizens expect national governments to deal with defense, foreign affairs, Social Security, environmental protection, and national projects such as space exploration or medical research. National governments seem further away to many citizens, and there is a greater degree of mistrust as well as a poorer understanding of what national governments do.

Overlapping areas of control for local and national governments include education, transportation, and commerce. Conflicting opinions of what local and national governments should do complicate decision making. In the United States, police are local, with only a few centralized agencies such as the Federal Bureau of Investigation, whereas in France police are organized nationally. Those who believe in strong national governments argue that centralized planning can be more effective, uniform, and efficient, in part because more experienced professionals will make better decisions based on more information. Those who believe in strong local governments claim that decentralized services are more effective because local officials are more in touch with citizen needs and can craft plans that are more in harmony with local situations.

Information and communication technologies have changed governmental decision making enough to require a reexamination of the balance of services. Centralization advocates argue that networked communications have made it easier for national government officials to be better informed about what is happening in local communities and to be responsive to local needs; therefore greater control should be given to central planners. Decentralization enthusiasts claim that new technologies have made them more efficient and sophisticated so that they can be more effective. They also argue that because of interconnectivity among local governments, uniform laws are easier to maintain thereby reducing the need for a central authority.

It appears that technology could be used to support either centralization or decentralization. Therefore resolution depends on the argumentation skills of individuals and the resources of political parties that are involved in the deliberation process. The latest corruption scandal or reform movement can shift welfare or healthcare into more centralized or decentralized forms.

For individuals, families, neighbors, colleagues, and citizens to get what they want from government, first they need to know what each of their governments is offering. The governments for the fifty United States or the twelve Canadian provinces have Web portals announcing their services. Washington state's award winning portal, Access Washington, <http://access.wa.gov>, features direct links to the state's most heavily used government Web areas and a powerful assembly of task-oriented tools for citizens and businesses (figure 9.1)[1]:

Jobs/work	Recreation	Visit Washington
State agencies	Services index	E-mail lists
Vital records	Resources list	Governor
Lottery results	Consumer help	Legislature
Licenses	WA Law search	Courts
Traffic watch	State facts	

9.1   Washington State Web page (part), <http://access.wa.gov/>. These materials have been reproduced with the permission of the Washington State Department of Information Services. © 1999–2001. All rights reserved. In addition, Access Washington™, Ask George™, and WaWizQuiz™ are trademarks and service marks owned by the Department of Information Services.

A separate area offers state emergency information for drought, fire, energy, and earthquakes. A closer examination yields seventy Web services directly for citizens, thirty-eight for businesses, and fifty-three for government agencies. Citizen services include vital records for life cycle events such as birth, death, marriage, and divorce; legislative information; and a low-income home energy assistance program. Future online citizen services will include vehicle and driver's license renewals and electrical permits. This is not your futurist's ream of a high-tech future but a steady progress of improving government-to-citizen services by facilitating access, reducing costs, and saving time. Unfortunately, Washington State's home page contains 269K bytes, by far the most of the fifty states, thus taking the longest to view.

Improving citizen-to-government communication has been the focus of several projects, such as the early Santa Monica, California, Public Electronic Network (PEN) (Van Tassel 1994). This well-off community had a high proportion of e-mail users even in 1990, prompting activists to initiate public information services. Citizen e-mail to city officials was generally a success, with citizens gaining satisfactory responses to their inquiries without substantially raising the workload of city administrators. Its early successes included the often retold story of how the PEN Action Group organized to help the homeless people who slept on the beaches (Varley 1991). The SWASHLOCK Project provided basics such as hot meals, showers, clothing, washing machines, and lockers, and then coordination with police, employment opportunities, and housing authorities. E-mail access was seen as the essential catalyst, even though there were problems with angry and mean-spirited messages. The homeless services were effective, leading some citizens to complain that it brought more homeless visitors. After a decade, Santa Monica now offers a wide range of Web services:[2]

Big Blue Bus	Engineering Division	Jobs
City Council Netcast	Environmental Programs	Library
City Council	Explore Santa Monica	Maintenance Management
City Hall Hours	Farmers' Markets	Municipal Code
City Services	Feedback to City Hall	Planning
CityTV	Find a Business	Police

Community Events	Fire Department	Recycling
Consumer Affairs	Forms On-line	Rent Control
Cultural Affairs	GIS	Seascape
Employment Opportunities	Human Services	Telephone Directory

Another famous success story using existing technology is the nongovernmental Seattle Community Network, where the crusading efforts of Doug Schuler and others have created an effective and extensive civic community in a major city.[3] More technically sophisticated projects to provide videoconferencing or voice-over IP (telephone connection by way of the Internet) enabled citizens who could not find or understand what they were looking at in Web pages to simply press a button to get a helpful state employee, who would walk them through the Web pages. The technology is appealing, but the substantial continuing state employee costs raise questions. However, if these state employees were involved in the continuing redesign of the Web site, they might provide valuable input to reduce costs by building more effective interfaces.

An activities and relationships table (table 9.1) shows that current government services fall largely in the first two columns: information provision with some opportunity for communication with government officials. New opportunities exist to support citizens who wish to propose new legislation, to form a public interest group, or even to create their own vacation plan. Then these activist citizens should be able to use Web sites to announce their community bake sale, parent-teacher association meeting, or high school football game. It would be a natural evolution for government to do more to enable citizen efforts in the "create"-"donate" columns.

The breakthrough idea is to apply technology in ways that serve human needs and have a huge impact on people's lives. Imagine if organizers from the U.S. Conference of Mayors, representing 1,100 cities with populations of 30,000 or more, coordinated their efforts.[4] The best features from current city Web sites could be included in a template whose common format would become familiar to citizens over time. The Santa Monica list of thirty topics could be expanded and organized in meaningful ways, translated into multiple languages, and automatically validated for completeness.[5] Then citizens could more reliably find what

A R T	COLLECT Information	RELATE Communication	CREATE Innovation	DONATE Dissemination
Self	Health, legal, and education info Unemployment and jobs	Protect privacy and free speech Ask for official help		
Family and friends	Economic info Travel and recreation Consumer affairs	Deliberations and petitions Consumer complaints		
Colleagues and neighbors	Local services Business and farm data Economic development	Form public interest groups	Form local and national proposals	Share solutions to civil problems
Citizens and markets	Laws and regulations Tax rules and forms Licenses and permits	Lobby for bills and regulations Solicit volunteers and campaign funds	Write new bills and regulations Organize political movments	Announce laws and regulations

TABLE 9.1   E-Government in an Activities and Relationships Table

they were looking for and search across all cities to compare healthcare or recreation services. Community organizers could find the most successful examples of public transportation, and business planners could compare corporate amenities and taxes. Similar strategies could be applied to the 12,000 U.S. cities with populations of 2,500 or more, or the 3,141 counties nationwide.[6]

Now imagine if similar ideas were developed to suit the needs of the 80,000 schools in the United States. Exemplary designs could be emulated so that each Web site would present information about the administrators and teachers, announce theatrical productions, science fairs, and sports events, and invite discussions among parents, administrators, teachers, and students. You could see what the curriculum was for your child, and how your school was performing compared to others in your state or nationally. These sites could also provide templates and processes for school districts to improve education, serve community needs, and organize local events. Then these Web sites could also facilitate dissemination so neighbors would know about the need for orchestral instrument donations, crossing guard volunteers, and parents to accompany students on a class trip.

These civic Web sites could help build cooperation and trust that support stronger schools and communities. Putting up the technology is only the first step; community involvement and enthusiasm are necessary. This is just the kind of application for which the Internet's distributed local nature fits perfectly—local control, effort, and empowerment, multiplied by 80,000 communities. Such Web sites might be a step in restoring some of the lost social capital, the willingness of people to participate in community activities, which has declined precipitously in the United States since 1965 (Putnam 2000). Active proposals include ideas for a Public Telecommunications Service, like the Public Broadcasting Service, that would provide a safe civic space for diverse creative public works projects and a safe place to establish durable human bonds among citizens (Levine 2000). A Public Telecommunications Service would be free of commercials, support open discussion of controversial issues, and foster community projects such as new parks. Technology volunteers, neighborhood associations, or a Civilian Information Corps (like the Civilian Conservation Corps of the 1930s) might be organized to help implement such services.

Other likely directions for future development are government-to-government services that could do much to improve efficiency. In our work with Maryland state agencies we were troubled to see that simple processes such as a request for a report on a juvenile offender from a school might take two weeks. The caseworker at the Department of Juvenile Justice who needed the report would ask the secretarial staff to prepare the proper request form for internal approval. Then the request would be mailed to the school, where it would be put in the stream of such requests for a response when time permitted. Since there were privacy concerns, careful preparation of the form was important. After logging the request, the online database was consulted to collect the requested data. Then the response was reviewed by a superior and mailed back to the Deptartment of Juvenile Justice, where the relevant information was entered into a database for viewing by the caseworker. Of course, many delays or errors can intrude such an unwieldy, slow, error-prone, and costly process. The obvious solution would be to establish a secure connection between the Department of Juvenile Justice, and the school district databases, but incompatible computer systems, differing terminologies, and varied administrative rules make this a substantial challenge. Businesses have often accomplished such integration, but governments have been slower to react because of scarcer resources, administrative barriers, and insufficient motivation. Accelerating such efficiencies would bring dramatic improvements and cost savings (Fountain 2001).

Government efforts to improve procurement could also bring better service and quality at lower cost. One estimate is that 20 percent of the annual $600 billion that the U.S. government spends on purchasing could be saved by reduced administrative costs and more competitive practices (Fountain 2001, 4). However, Jane Fountain warns that "the initial euphoria in the public that greeted e-commerce has been replaced with a growing awareness of the painstaking, painful organizational and industry restructuring that will be necessary. . . . Government is following a similar trajectory . . . [but it] is far more difficult."

Ironically, one of the most complex but rapidly growing government services is electronic tax filing. During 2001 more than 30 percent of the tax forms in the United States were filed electronically. No one wants to pay taxes, but making the process fast so that refunds are received promptly attracts many people. Tax filing has been accelerated by the highly effective tax preparation software tools that do an excellent job of guiding users through an extremely complex process. Intuit's TurboTax and its competitors, such as H&R Block's TaxCut, are excellent examples of good designs that structure complex processes with comprehensible overviews and empower users by allowing them to see the impact of their changes.[7]

GETTING THE GOVERNMENT YOU WANT

The naive wish to elect the candidates you want and get government to do what you want gives way quickly to several disturbing realities. Your candidate may not be the majority candidate, your representatives may pass laws you don't like, and your government may have limited resources. Getting what you want from modern democracies is not easy because it requires one of society's more elusive notions— consensus of your fellow citizens or their representatives. Since consensus building is essential, new technologies that promote communication and collaboration will have a substantial influence on democratic processes.

I got a lesson in the difficulties of consensus building while participating in the public policy group of my major professional society, the 85,000-member Association for Computing Machinery. After much internal discussion and consensus building about how to make our point clear, we carefully drafted a letter to promote a U.S. congressional bill on privacy protection. I thought that a well-crafted letter from such a society's respected president would be seen as strong support, but I quickly learned that our society was considered only a modest-sized group and that congressional staffers would want to know if other scientific societies shared our enthusiasm for the bill. To handle such coordination there is a Council of Scientific Society Presidents in Washington, D.C., that draws participation from seventy-five leading professional associations. Our group sent the proposed four-paragraph letter to this group's listserv asking for supporting signatures within four days. These other scientific societies convened their boards of directors or policymakers and by the following day we began to get the supporting notes we wanted, but then on the next day several societies objected to our third paragraph and indicated they would sign on only if we changed its wording. Given the approvals we had already gotten and the limited time frame, we could not agree to their changes and lost their support, thereby reducing the impact of our letter.

Building consensus is tough, and doing it quickly is even more difficult. Compromise takes time, and that is one reason why democratic governments often change slowly, thus preserving the status quo. Information and communication technologies may speed up the process of information gathering and discussion, but it is equally effective for supporters and opponents of a change. While compromise may take just as long in the digital age, we can hope that each side is better informed and has more refined proposals.

Commentators debate whether consensus building, or even consensus seeking, is easier face-to-face or online. A common first reaction is that online discussions are inferior to face-to-face ones, which are the gold standard to which designers should aspire. Face-to-face discussions enable subtle emotional expression and reactions, allow quick interchanges, and have the compelling potency of physical presence. However, online discussions can be more inclusive, enabling participation by those who might not be able to travel or who feel uncomfortable speaking in public. Online participants can send their comments at any hour of the day and can calmly reflect on what they read and write. Empirical evidence shows more democratic participation and diverse points of view in online discussions as compared to face-to-face encounters. However, there is no need to choose between face-to-face and online discussions. Both are helpful and complement each other very productively. Face-to-face discussions can be valuable trust-building encounters that improve online discussions. Online discussions can supplement face-to-face meetings by allowing position papers to be sent in advance and fostering follow-up discussions on missing information or dropped arguments.

A natural first step for getting involved in online democratic processes would be to promote legislation in an existing parliamentary body such as a town council or state legislature. One might begin by gathering support among neighbors or political party members to nominate candidates and then promote their election among fellow citizens.

The next generation of software tools could do much more than e-mail to support these processes.

Let's start with a simple example to see how your capacity to collect signatures from your neighbors for getting a traffic signal at a busy school crossing could be enhanced by the next generation of well-designed e-government tools. First you would find out how decisions are made about traffic signals in your community. Your local traffic engineer would guide you to the signal warrants in the Manual of Uniform Traffic Control Devices. Then you would collect current information (from sensors or satellites) on automobile and pedestrian traffic rates, with a detailed history of the growing number of accidents at that school crossing (from properly annotated police and hospital logs tied to a unified geographic information system). You would document the number of similar school crossings that already have a traffic signal and show reduced accident rates. Next you might prepare a report as an attachment to a draft petition to the proper office of your local highway authority, using the template it provides on its Web site. Now you are ready to circulate the draft to your neighbors for comments using the community e-mail list derived from voting records, all with proper privacy protection. After a week of comments, you would revise the petition according to your neighbors' suggestions and request electronic signatures on the petition, which you would send to the appropriate official with a request for a response within thirty days.

This happy scenario can become reality among family, friends, close neighbors, and trusted colleagues without too many technology advances. However, even in this safer circle of relationships, problems would emerge if some neighbors objected to the traffic signal because they honestly believed it would slow down traffic, increase congestion, or cost too much to install. Then your request for signatures would run into problems, and a neighborhood debate could boil over into angry discussions. Strong disagreements occur daily over neighborhood issues such as cutting down trees, building ugly garages, or playing music too loudly.

When city or state government decisions must be made among larger groups of less familiar participants, more structure and support are needed. When national political discussion is required for more volatile issues such as abortion, handgun control, welfare, or environmental protection, the complexity grows because well-organized groups with out-spoken leaders and substantial funding become active participants. As new issues emerge, such as monitoring e-mail or bank transfers to detect terrorist activity, established groups have an advantage because of their reputation, but new organizations can focus on current topics with fresh perspectives. However, the open nature of the Internet means that producing consensus may not be any easier in the online world than in the real world.

The necessary advances in technology might be reached by thoughtful application of new computing tools that help to create consensus. The goal is open deliberation within small established communities (under one hundred people) and also in larger communities (more than one thousand people) in which there is less shared knowledge and trust. By forming small public interest groups of like-minded citizens with whom you develop an understanding and high level of trust, you can begin to assemble the influence you will need to wage campaigns for national issues and candidates to represent you in local and national governments.

OPEN DELIBERATION

Political philosophers have long debated the nature of open deliberation under many terms: unfettered discussion groups, lively café society, the public sphere, rational debates, respectful dialogues, civic spaces, community meetings, sounding boards, public forums, town halls, round tables, democratic salons, city councils, popular parliaments, and various forms of commons, senates, assemblies, committees, agencies, congregations, conferences, conventions, courts, boards, and commissions. The rich language and diverse metaphors suggest the pervasiveness of the desire for open deliberation well before the appearance of the Internet. Deliberation is meant to convey the combination of collaboration and competition, and the back-and-forth dialogue that leads to compromises, refinements, and amendments that mark the road to consensus.

Open deliberation is more than having everyone speak their mind. A torrent of e-mail postings in a large group is not a discussion and may never converge to a consensus. Many models of open deliberation are possible, but let's start by adopting existing parliamentary processes. A recognized delegate might propose some action (make a motion for adoption) by e-mail in a clearly stated manner so everyone understands what is at issue, for example, making abortion or handgun ownership illegal. Proposals and amendments might be sequenced and put up for time-limited online debate followed by electronic votes. Comments could be reviewed to ensure that they were focused on the topic, limited in length, and respectful to others.

One of the key concerns about electronic debates is that they should integrate the comments of previous messages and ask clarifying questions where appropriate. Evidence from some analyses shows that only a fraction respond to previous messages; everyone may promote their point of view without acknowledging what others have said, and therefore they may be slow to compromise. Critics argue that face-to-face debates may be more likely to produce compromise and changes in positions because of the stronger emotional presence and the sequential nature of debate. On the other hand, the typewritten approach to online debate may cause arguments to be presented more carefully and less emotionally while permitting the readers to review and reflect before responding.

Designs for open deliberation might include forced delays in responding, limitations on how often each person can submit messages, and requirements to respond to existing proposals before presenting new ones. Straw poll methods to get a sense of the group's opinions could accelerate discussion on issues for which there is strong agreement, and confusion indicators could allow participants to indicate when the arguments have gotten so complex that a clarification summary is needed. The design possibilities are vast.

Promoters of each new medium see an opportunity to support open deliberation, such as a newspaper's letters to the editor, talk radio, and public access cable TV. However, the scarce resources of linear time and limited bandwidth coupled with the centralized broadcast nature of most technologies limits open deliberation. The Internet has a particularly rich set of support tools for open deliberation because of its fundamentally decentralized structure, which can give each user a voice. Maybe not an equal voice, but at least a voice.

While postal mail has an appealing physical presence, e-mail is potent because of its capacity to reach many people rapidly and cheaply. Newsgroups and Web-based threaded discussion groups are less intrusive than e-mail because the notes are simply posted to a public location, requiring recipients to visit when they remember to make the effort. These technologies are *asynchronous*—one can access notes any time of the day or night, whenever it is convenient. This asynchrony also allows participants to reflect on what they have read, consider their responses, and prepare them carefully.

Another family of electronic technologies for open deliberation is the *synchronous* chat, which requires the simultaneous presence of participants, thus severely limiting the capability for reflective commentary. Synchronous chats typically contain one-line comments, hide the identities of participants, and rarely archive discussions.

The good news about all these technologies is that they are relatively low-cost and fairly easy to learn. Their textual content requires only minimal computer power, and therefore they are accessible to low-income or disabled users. The text can be read aloud to vision-impaired users or to those desiring telephone access, and roughly translated automatically for foreign language speakers. The plasticity of textual messages means that users with small wireless portable devices can read messages anywhere.

Jenny Preece's framework for promoting sociability in online communities (see chapter 8)—people, purposes, and policies—will help establish open deliberation. You will want to create a well-defined community of users, clearly state your purpose, and provide simple policies about staying on topic and avoiding harsh language. If you have a clearly stated purpose, such as a petition to install a traffic signal on a busy school corner, a small number of neighbors is all that you need, and discussion policies can be informal.

In shifting to open deliberation for national political debate, you may have additional worries. The central concerns are the stronger requirement for inclusiveness or universality, and the accompanying need to scale up to accommodate millions of users. Medical and political groups can be useful with dozens or hundreds of users, but substantive political debate on national issues must eventually include millions of diverse citizens.

Andrew O'Baoill (2000) applies the political philosophy of Jurgen Habermas (1989) to analyzing public debate in Internet forums such as Slashdot.[8] Habermas celebrates multidisciplinary perspectives to understand political processes and idealizes eighteenth-century coffeehouses and literary salons as intellectual forums where ideas were exchanged in respectful face-to-face discussions. O'Baoill interprets Habermas's writing about the "public sphere" that forms public opinion through

> *Universal access.* Anybody can have access to the space.
> *Rational debate.* Any topic can be raised by any participant, and it will be debated rationally until consensus is achieved.
> *Disregard of rank.* The status of participants is ignored.

These ideas suggest useful principles to guide design of online discussions. Anthony Wilhelm (2000) pursues these themes to develop what he calls the topography of the virtual political public sphere:

> *Antecedent resources*—the skills and capacities that one brings to the table
> *Inclusiveness*—ensuring that everybody affected by a certain policy has the opportunity to access and use essential digital media
> *Deliberation*—subjecting one's opinions to public scrutiny for validation
> *Design*—the architecture of the network developed to facilitate or inhibit public communication

Habermas's universal access and rational debate become Wilhelm's inclusiveness and deliberation. However, Wilhelm's antecedent resources raise key universal usability issues that are more troubling in online than face-to-face discussion. Online participants must have technological capacity and the skills to use it, which is a much higher requirement than for face-to-face participants. This is a major concern that demands substantial attention and investment. For other applications such as e-learning, e-business, and e-healthcare, the more people who participate the better, with benefits growing in proportion to the number of people who participate. However, if societies are to apply online political discussions that create public consensus to guide government action, there is a strong requirement for universal participation. Satisfying this requirement will require more than low-cost services; it will promote the case for improved training and better user support services.

Wilhelm's design theme brings us back to the impact of design on deliberation—both the usability and the sociability. Synchronous chat groups tend to be rapid-fire and far from the reflective deliberation that supports consensus building. Asynchronous threaded discussions are more likely to support the back and forth argumentation that clarifies intentions, refines proposals, and enables participants to change their position. Respectful exchanges open the door to consensus.

The likelihood that broadband communications will somehow remove the problems is dubious, nor will immersive three-dimensional virtual realities save the day. These technological advances may be useful for some activities, but the stronger need is for tools to support facilitation, moderation and mediation while limiting harmful activities such as angry flaming (hostile and inflammatory notes), off-topic discussions, and disruptive behavior. In short, we need a modern online Robert's Rules of Order. This 1870s guidebook for conducting meetings laid out the procedures and etiquette for respectful debate and fair voting, and provided procedures to handle disagreements and disruption. Our goal should not be to replicate the real world but to create a better than real-world experience.

Online communities have begun to establish written policies that describe permissible behavior and dispute resolution. As these policies are applied for political discussions (figure 9.2), business committees, and other governing bodies, refined versions with clear statements of bylaws, procedures for voting, and formats for recording decisions are emerging. Among programmers, the lively and informed discussions on Slashdot ("News for Nerds. Stuff That Matters") generate vigorous debate and "flame wars," but rules of moderation and metamoderation keep discussions under control.

Fair and clear rules of behavior create a safe place for open deliberation, where responsibility for actions coupled with a history of reciprocity generate trust and encourage cooperation. Institutionalizing these rules and embedding them in software will take decades, but these are necessary steps to raising online communities from an appealing forum for many to a reliable constructive political process for all.

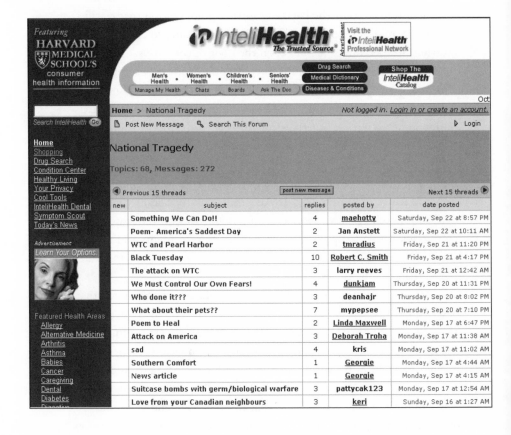

9.2 Community Board at <www.Intelihealth.com>. Screenshot courtesy of Intelihealth, Inc. (an Aetna company).

Some critics of politics on the Internet worry that discussion groups generate only participation by like-minded individuals who merely reinforce their own beliefs online because the deliberation is less than open (Sunstein 2001). While narrowly focused groups are likely to emerge, this seems to reflect the real world, in which people are drawn to spend time with people like themselves. This fragmentation builds local and topic-oriented groups of many sorts, but they will eventually have to engage in open deliberation with those holding opposing views.

To support a national consensus that reflects the views of millions of citizens through a mix of representation and direct participation will take ambitious development. Even basics such as setting an agenda, limiting discussion, caucusing in smaller groups or subcommittees, and posting public summaries will take innovative design and refinement through testing.

And yet, these foundational steps are the easy part. More serious challenges will come when dealing with powerful influences such as large corporations that promote their causes vigorously or military leaders who use force to override group decisions. Traditional challenges to democratic processes will also emerge in online politics, such as racial bias that seeks to limit one group's participation or factional disputes that could become violent. Online political debates will also be the target of mean-spirited hackers who use the technology maliciously to disrupt discussion, much like screaming protesters or filibusterers who prevent others from speaking. More subtle interference could include diverting messages from certain participants or tampering with votes subtly enough to shift the balance while avoiding suspicion. Democracy has never been easy, and it won't be easy in cyberspace.

Political parties seem likely to continue increasing their use of the Internet for campaigning and fund raising. The allure of reaching many voters at low cost with specific messages based on their age, gender, economic situation, or political outlook is very appealing to party officials. But it turns out that e-mail messages can look like annoying spam if they reach the wrong people, just like misdirected e-business mailings. Engaging voters in political issues will prove to be harder than expected, but volunteers who open up a dialogue with sympathetic voters may be more effective in winning their votes or getting contributions. Even more effective will be the use of e-mail to organize the party faithful and conduct discussions among convention delegates.

Another beneficiary of e-mail use will be public interest groups, which can more easily organize potential members, engage volunteers, and solicit contributions. Whether you are for environmental protection or oil drilling, for free markets or fair markets, or for animal rights or fur coats, you will be able to find or form a group of like-minded individuals.

A special case of democratic processes is the regulatory hearing held by many U.S. government agencies such as the Federal Trade Commission or the Federal Communications Commission (Shulman et al. 2001). These agencies solicit comments from individuals, citizen groups, industry leaders, and professional associations about specific issues or decisions. These public forums, which are typically held in Washington, D.C., or selected U.S. cities, have begun to include online components. Online forums enable those who cannot travel to the live forum to contribute their suggestions. Some fear that easy access to these forums will raise costs and cause delays while each comment is analyzed, but the benefits of broader participation seem large and the costs can be contained.

THE SKEPTIC'S CORNER

Ah democracy! The pundits (starting with Winston Churchill) say it is a terrible system but better than any other form of government. Ah bureaucracy! Incompetent, unmotivated civil servants with inadequate technology and limited resources, preventing respectful citizens from getting what they deserve. The skeptics believe that government reform is hopeless and that the best direction is less government. The engaged activists hope that cleaning up political contributions and hiring professional managers will invigorate politics and make governments more efficient.

Politics will never be completely free of undue influences, but open deliberation can give more people a voice and provide more wise proposals. Limiting the influence of powerful business, military, or religious factions won't be easy because these well-organized and powerful forces can exert substantial pressure. And there are still more dangerous disruptive forces such as corrupt officials, deceptive leaders, violent extremists, and deadly terrorists. These antidemocratic forces won't be bothered by gentle intellectual debates; they will remain a challenge unless open deliberation and processes can mobilize sufficient public opinion and equally powerful counterforces.

Government can never give all citizens what they want, but it can be made to function more effectively. Technology can support improved efficiency in government processes, but the major salutary effects are likely to be the empowerment of individuals to be vigorous in their pursuit of getting what they want from government and their capacity to build consensus through open deliberation. Politics has been called the art of the possible, but sometimes technology can change what is possible.

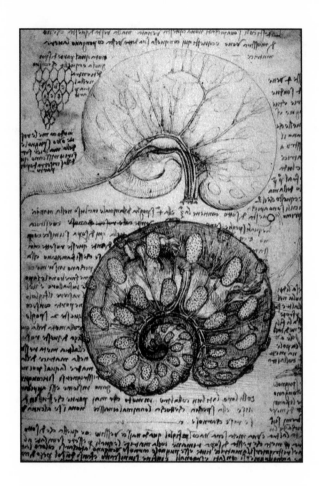

Leonardo's attempt to find similarities among natural forms, from license-free "Leonardo da Vinci: Selected Works," Planet Art.

10

Creativity can solve almost any problem. The creative act, the defeat of habit by originality, overcomes everything.
—George Lois

LEONARDO'S CREATIVITY

Controversies over Leonardo have raged over the past five centuries, but everyone agrees that he was creative. His work spanned many disciplines, although during his lifetime he was known mostly for his art. Leonardo's reputation as a scientist grew after his death as scholars studied his notebooks and marveled at his diverse accomplishments. The thousands of pages of his secret writing reveal his explorations and inventions in geology, astronomy, optics, hydraulics, and aerodynamics. They also present his anatomical drawings from secretive night-time dissections of cadavers and animals. Again and again, commentators wonder about how the history of science might have been accelerated by open presentation of Leonardo's insights and conjectures about geology, telescopes, or blood circulation.

Freud (1910) claimed that Leonardo became creative because of his sublimated sexuality and to overcome his rejection as an illegitimate child. But there is good reason to think that Leonardo was loved and cared for as a child by his mother, and then by his grandparents, father, stepmother, and numerous stepbrothers and stepsisters. His father encouraged and acknowledged his artistic skills, apprenticing him as a teenager to the great Florentine artist Verrocchio. The studio community at Verrocchio seems to have had a strong influence because Leonardo always formed a circle of people around him. He had scholar-colleagues such as famed mathematician Luca Pacioli and younger assistants such as Andrea Salai and Francesco Melzi. Freud also described Leonardo's homosexuality, but little evidence of his relationships with men or women is based on historical evidence. Leonardo seems asexual and unemotional but intensely devoted to his explorations.

We can all learn something from Leonardo. His thirst for understanding diverse knowledge domains and willingness to take on ambitious projects are good inspiration for those who aspire to be creative.

This chapter reviews three schools of creativity and offers a vision of how software could enable more people to be creative more of the time. The term *mega-creativity* is meant to convey the idea that millions of people could benefit from such creativity support tools.

INSPIRATIONALISTS, STRUCTURALISTS, AND SITUATIONALISTS

The large literature on creativity offers diverse perspectives on what it is and how to get it. Some writers, I'll call them *inspirationalists,* emphasize the remarkable "Aha!" moments in which dramatic breakthroughs magically appear. A famous legend depicts Archimedes (third century B.C.) jumping from his bath screaming "Eureka!" as he realized that measuring the amount of water displaced by the king's crown would indicate if it were made of gold. Another classic story that emphasizes the intuitive aspects of creativity tells of a dream of Friedrich August Kekulé (1829–1896) about a snake biting its tale. This circular vision was said to have led him to discover the ring structure of benzene.

Most inspirationalists are also quick to point out that "luck favors the prepared mind," thereby turning to the study of how preparation and incubation lead to moments of illumination. The inspirationalists also recognize that creative work starts with problem formulation and ends with evaluation plus refinement. They acknowledge Thomas Edison's (1847–1931) balance of 1 percent inspiration and 99 percent perspiration—a flash of insight followed by much hard work to turn it into a practical result.

Those who emphasize this inspirational model promote techniques for brainstorming, free association, and divergence. They advocate strategies to break an innovator's existing mindset and have him perceive the problem with fresh eyes. Since they want innovators to break from familiar solutions, their recommendations include travel to exotic destinations with towering mountains or peaceful waterfalls. Inspirationalists talk about gifted individuals but usually stress that creativity-inducing thought processes can be taught.

The playful nature of creativity means that software support for inspirationalists emphasizes free association using textual or graphical prompts to elicit novel ideas. Random words, rapid flipping through photos, or abstract shapes like inkblots have all been suggested as stimuli for creativity. Leonardo understood the benefits of random visual imagery as a source of inspiration. He recommended studying "stains on walls or embers of a fire wherein we may find divine landscapes, battles, figures in violent action or even the expressions of faces."

Inspirationalists are often oriented to visual techniques for presenting relationships and for perceiving solutions. They would be sympathetic to information and scientific visualization strategies that help you understand previous work and explore potential solutions. Many writers and software developers of tools such as IdeaFisher or Mind-Mapper encourage two-dimensional layouts of loosely connected concept nodes to avoid a linear or hierarchical structure. They encourage you to adopt a casual style and delay judgment about new ideas. They want you to use informal sketches that are easy to change or toss out. Inspirationalists would also appreciate templates as starting points for a creative leap, as long as powerful tools enable you to explore fresh combinations.

A second group of writers on creativity, the *structuralists,* emphasizes more orderly approaches. Structuralists stress the importance of studying previous work and using methodical techniques to explore the possible solutions exhaustively. When you find a promising solution, you should evaluate its strengths and weaknesses, compare it to existing solutions, and refine the promising solution to make it implementable. Structuralists teach orderly methods of problem solving, such as George Polya's four steps in his 1957 classic *How to Solve It:*

1. Understanding the problem
2. Devising a plan
3. Carrying out the plan
4. Looking back

For structuralists, libraries and Web sites of previous work are important, but the key software support comes in the form of spreadsheets, programmable simulations, and scientific, engineering, or mathematical models. These software tools support what-if processes that allow you to try out your assumptions. You can see the slowdown in traffic if one lane is blocked or the rise in sea level if you add more carbon dioxide to the atmosphere.

Structuralists often encourage you to show processes with visual animations, such as the functioning of heart valves or growth of crystals. Structuralists are usually visual thinkers, but instead of sketches they prefer orderly flow charts, precise decision trees, and structured diagrams. Since they favor methodical techniques, they are likely to appreciate software support for step-by-step exploration, with the chance to go back, make changes, and try again.

A third group, the *situationalists,* emphasizes intellectual, social, and emotional contexts as key parts of the creative process. They see creativity as embedded in a community of practice with changing standards, requiring a social process for discussion and approval. Scientists and physicians have to send their papers for review to scientific journal editors before publication. Artists have to submit their paintings to gallery curators, and writers have to wait to hear from literary prize juries.

The famed University of Chicago psychology professor Mihaly Csikszentmihalyi (1996) is a key thinker among the situationalists. His research and interviews with famously creative people led him to identify three components of creativity:

> Domain, such as mathematics or biology, which "consists of a set of symbols, rules and procedures"

> Field, which "includes all the individuals who act as gatekeepers to the domain. It is their job to decide whether a new idea, performance, or product should be included in the domain."

> Individual person, whose creativity is manifest "when a person using the symbols of a given domain such as music, engineering, business, or mathematics has a new idea or sees a new pattern, and when this novelty is selected by the appropriate field for inclusion in the relevant domain"

Csikszentmihalyi writes thoughtfully about the motivations of individuals and their desires to impress members of the field and to expand the domain of knowledge, or better still, to start a new one.

Situationalists are most likely to talk about the influence of family, teachers, peers, and mentors. They want to understand what kind of childhood experiences, such as parents who encouraged question asking or sibling rivalry, motivated a creative life. They consider the influence of challenges from memorable teachers and the strong desire to create because of its inherent satisfaction. Situationalists also explore why individuals pursue recognition and how they overcome failures, even in the face of competition. They want to know about the role of collaboration with peers, advice from mentors, and emotional support from a spouse or friend (Gardner 1993).

For situationalists, vital user interfaces are those that support consultation with peers and mentors and dissemination of results to interested members of the field. They are eager for access to previous work in the domain by way of Web search engines and digital libraries. They believe in the importance of reward for accomplishment and fear of failure as motivators. Public prizes motivated John Harrison to solve the problem of longitude by building an accurate chronometer in the eighteenth century and prompted Lindbergh to fly solo across the Atlantic Ocean in 1927.

These three perspectives on creativity—inspirational, structural, and situational—are all useful creative strategies whether you use software or not. But in choosing software tools, each perspective leads to different outcomes. If you have an inspirationalist leaning, you can choose tools that provide random image libraries, link to associated ideas, and offer templates for initiating action. If you want to follow a structuralist path, you can exhaustively explore alternatives and use simulation models to understand the impact of each choice. If you have the social instincts of the situationalists, you can emphasize consultation by e-mail and more refined methods to support your creative endeavors.

THREE LEVELS OF CREATIVITY: EVERYDAY, EVOLUTIONARY, REVOLUTIONARY

Only some knowledge work or art products deserve to be labeled creative. Many people's work is merely repetitive application of rules, such as deciding on mortgage loans or finding a flight reservation from New York to London. Some jobs require competent original work, such as making a financial plan or planning travel itineraries through Europe. Higher levels of creative work might involve finding patterns of loan defaults from a huge database or opening up new tourist destinations.

Similarly, a politician's speech often has rote-memorized phrases, but usually it contains original sentences and sometimes brilliant creative comments. Inspired lectures and creative rhetoric, such as Martin Luther King's "I Have a Dream" speech are rare. Similarly, redrawing a travel map to your home is copying, doodling on an envelope may be original, but Picasso's drawings in the Vollard Suite reach the level of creative work. The proposals in this chapter are intended to support creative, not merely original, work.

The large literature on creativity considers diverse levels of aspiration. A low-level definition of creativity would include *everyday* impromptu or personal creativity. Can your lively conversation or attentive parenting be considered a part of the creativity spectrum? These more spontaneous and private activities may be creative in a broad sense, but since they seem harder to support and evaluate, let's put them aside for this discussion.

A familiar definition of creativity would include what Thomas Kuhn in his classic book *The Structure of Scientific Revolutions* (1962) referred to as normal science. He was describing useful *evolutionary* contributions that refine and apply existing paradigms or methods of research. Evolutionary acts of creativity include doctors making cancer diagnoses, lawyers preparing briefs, or photo editors producing magazine stories. Their work is important in changing someone's life by medical care, legal practice, or journalistic reportage. It also satisfies another criterion of creativity: that the product be public so that others can assess it.

Evolutionary creativity is the focus of this chapter, in part because it is most likely to be helped by software tools. There is a chance that software tools that support you in evolutionary creativity may also help you produce revolutionary breakthroughs. On the other hand, it is possible that software tools that support evolutionary creativity restrict your thinking or even discourage paradigm shifts. If your software lets you do amazing things by manipulating still photos, you may never think of doing animations or writing music.

While most people talk about their everyday creative acts and some people can report on their evolutionary creative contributions, few people can claim revolutionary creativity breakthroughs. The restrictive definition of *revolutionary* creativity focuses on great breakthroughs and paradigm-shifting innovations, as described by Thomas Kuhn. Einstein's relativity theory, Watson and Crick's discovery of DNA's double helix, or Stravinsky's *Rite of Spring* are often cited as major creative events. Such a definition confines our discussion to rare revolutionary events and a small number of Nobel prize candidates. It would be difficult to design support tools for such revolutionary thoughts because they are by definition breaks with the past and therefore unpredictable.

Therefore our focus is not on everyday or revolutionary creativity but on the middle ground: evolutionary creativity. This still covers a wide range of possibilities. Developing, finding and using software support tools for evolutionary creativity according to the three perspectives—inspirationalist, structuralist, and situationalist—is a sufficient challenge. My goal is mega-creativity—to enable more people to be more creative more of the time.

A FRAMEWORK FOR MEGA-CREATIVITY

This chapter attempts to define the powerful tools that can facilitate creative work by many people. It builds on the four activities of the framework developed in chapter 5:

COLLECT Learn from previous works stored in libraries, on the Web, and other places

RELATE Consult with peers and mentors at early, middle, and late stages

CREATE Explore, compose, evaluate possible solutions

DONATE Disseminate the results and contribute to libraries, the Web, and other places

These four activities are not a linear path. Creative work may require you to return to earlier phases and much iteration. For example, libraries, the Web, and other resources may be useful to you at every phase. Similarly, you may want to open discussion with peers and mentors repeatedly during the development of an idea. The social processes that support refinement can also be helpful to you at early, middle, and late stages of the creative process.

When you come up with something new and disseminate it to others, your work becomes publicly available. This makes it a candidate for the next person to build on and learn from your work. Personal computing technologies coupled with networking have made databases widely available. Patents, legal decisions, and scientific papers, as well as music, poetry, novels, art, and animations, are available online. We are still far short of having access to all of these materials because of financial limitations, copyright issues, and business models that seek to profit from creative work.

This framework has much in common with previous descriptions and methods, but there are important distinctions. Daniel Couger (1996), an information technology professor, reviewed twenty-two "creative problem solving methodologies" with simple plans such as

> Preparation
> Incubation
> Illumination
> Verification

as well as an even more basic, three-phase plan:

> Intelligence: recognize and analyze the problem
> Design: generate solutions
> Choice: select and implement

Couger offers his own plan with five phases:

> Opportunity, delineation, problem definition
> Compiling relevant information
> Generating ideas
> Evaluating, prioritizing ideas
> Developing an implementation plan

It is striking that so many of the plans limit themselves to the narrow perspective of the inspirationalists and structuralists. Problem solving and creativity are portrayed as lonely experiences of wrestling with a problem, breaking through various blocks, and finding clever solutions. The descriptions of even early phases rarely suggest contacting experts on the problem or exploring libraries for previous work. Consultation with others and dissemination of the results are minimally mentioned. Some reconsideration of such methodologies is in order because of the presence of the World Wide Web. It has already dramatically reduced the effort of finding previous work, contacting experts, consulting with peers and mentors, and disseminating solutions.

Of course, there are costs and dangers in reading previous work or consulting with others. There are also satisfactions in solving a problem on your own. But when dealing with difficult problems, the benefits of building on previous work and consulting with peers and mentors can be enormous. The framework here builds on the situationalists' perspective by embracing the expanded opportunities offered by the World Wide Web.

The goal of the framework is to suggest ways to use and improve Web-based services and personal computer software tools. With the reduction of distraction caused by poorly designed user interfaces, inconsistencies across applications, and unpredictable behavior, users' attention can be devoted to the task. In an effective design, the boundaries between applications and the burdens of data conversions would disappear. Data representations and available functions would be in harmony with problem-solving strategies. Then the user would be in control and have the sense of mastery that makes possible concentration on the four activities in creative work: collect, relate, create, and donate.

My suggestion of a close link between supporting creativity and generating excellence is a hope, not a certainty. Creativity support tools may be used to pursue excellence, high quality, and positive contributions, but I am sadly aware that this will not always be the case. The framework could tighten the linkage because of its emphasis on consultation and dissemination. I believe that making creativity more open and social through participatory processes will increase positive outcomes while reducing negative and unanticipated side effects.

Leonardo recognized the importance of collecting previous work as a basis for new work. He had a personal library of at least forty-five classic books and had access to several libraries, at a time when books were still rare. He discussed some of his explorations with others and learned what he needed from scholar friends such as mathematician Luca Pacioli. Leonardo noted others' ideas, often copying extensive passages into his notebooks. He carefully studied classical Greek and Roman sources but made some of his greatest contributions by rejecting accepted principles and offering his own analyses. He was bold in countering Aristotle's sympathy for syllogisms (logical statements) with a more scientific approach based on experimentation. He was courageous in questioning beliefs about the unchanging earth by suggesting that the Tuscan mountains were once under the ocean. Visual methods were more important than mathematics for Leonardo. His drawings of flying birds, light rays, or water flow showed a masterful understanding of what he saw, in ways that would be reduced to mathematical formulae more than two hundred years later. Of course, if he had shared his work, he may well have lost his prestige or been excommunicated.

Leonardo's example can teach us much about the importance of dissemination, but here the story is a cautionary one about how failure to disseminate limits the progress of others. Much of Leonardo's work was hidden from view and later destroyed. A historian of medicine (Nuland 2000) wrote, "Had he produced the anatomical textbook which he had planned . . . the progress of anatomy and physiology would have been advanced by centuries." What a loss for medicine. Arnold Hauser (1966) summarized that "his findings remained his alone; they were not contributed to the public fund of common knowledge and therefore not able to influence a community endeavor of assumed social importance" (4).

INTEGRATING CREATIVE ACTIVITIES

The creativity framework will work only if there is integration of multiple creativity support tools. Some of these tools already exist but could be enhanced to better support creativity. However, the main challenge for users and designers is to ensure smooth integration across these novel tools and with existing tools such as word processors, presentation graphics, e-mail, databases, spreadsheets, and Web browsers.

Smoother coordination across windows and better integration of tools seems possible. Just as word processors expanded to include images, tables, annotations, and more, the next generation of software is likely to integrate additional features. The first aspect of integration is data sharing and it can be accomplished simply by providing compatible data types and file formats. You should be able to import weather data from a Web page into a forecasting program so you can make your own prediction. You should be able to download a song and put it into a composition tool so that you can read the notes and make your own variation or use it as background for a video you are making. Of course, you might have to pay for the right to do this, and the cost could be added to your phone or Internet account bill.

A second aspect of integration has to do with compatible actions and consistent terminology. Most computer users are familiar with patterns of actions such as cut-copy-paste or open-save-close. Higher levels of actions that are closer to your task might be candidates for inclusion in the next generation of tools, such as annotate-consult-revise or collect-explore-visualize. Until these functions become available in standard tools, users will have to adopt careful working styles to make this possible.

For example, one devoted family photographer created photo collections with captions of who was in each photo and a record of the event (annotate). She sent these by e-mail attachments of a word processor file and twenty-five photo files to family members for their comments, reminiscences, and stories (consult). Then when they sent back their comments by e-mail, she deleted the least-liked photos and added the best comments (revise) into a final album that was archived as a set of Web pages. This was a pretty complicated task that took time and expertise beyond what most users are capable of doing. However, if there were a specific tool to support the annotate-consult-revise process for photo libraries, more people could consistently produce creative results that captured family personalities. Of course annotate-consult-revise, can also be applied to scientific papers, musical compositions, or architectural drawings.

Similarly, wouldn't you like to have a collect-explore-visualize tool set to let you gather family genealogies, sales information, or book citations? You could describe your needs, such as your family information, then invoke a search task (collect) to gather information from multiple Web sites and libraries. Next you would review the result sets, selecting some, rejecting others, and putting some aside for later review. Finally, you could visualize the results in a family tree, a historical time line, or a world map.

A third aspect of integration is the smooth coordination across windows. For example, if you see an unfamiliar term in a Web page, you should be able to click on it and get an English definition, a French translation, or a medical dictionary report, all in a predictable screen location. Similarly, if you find a personal name in a news report, you should be able to get a biography, e-mail address, or contact information. Such tools are available, so you can set up some of the services you want immediately. However, more ambitious tools could be even more helpful. If you have a map of your city, you should be able to click on landmarks and get explanatory Web sites, travel directions, or construction diagrams. If you select a region, you should be able to get the demographics of the population living there, a table of the entertainment events, or photos on a time line showing the history. Of course, the ambition can be even greater. You might want to click on one of the photos and get biographies of the people in the photo or the complete archive of the photographer. Smooth coordination enables you to rapidly pursue connections and lowers the barriers to creative activities.

Integration and smooth coordination will benefit many users for many tasks. The remainder of this chapter probes more deeply into specific tasks to support the collect-relate-create-donate activities. Adding the three perspectives—inspirationalist, structuralist, and situationalist—helps lead to useful suggestions. I propose eight specific tasks that should help more people be more creative more of the time:

> *Searching* and browsing digital libraries, the Web, and other resources
> *Visualizing* data and processes to understand and discover relationships
> *Consulting* with peers and mentors for intellectual and emotional support
> *Thinking* by free association to make new combinations of ideas
> *Exploring* solutions—what-if tools and simulation models
> *Composing* artifacts and performances step by step
> *Reviewing* and replaying session histories to support reflection
> *Disseminating* results to gain recognition and add to the searchable resources

I can't prove that these eight tasks are a complete set, but they can help as a checklist if you are looking for software tools or considering designing some new ones.

You can use existing general-purpose software tools to support these eight tasks, but developers are coming up with specially tailored products in your domain of work. Searching is a hot topic, and many researchers and companies are proposing improved search tools for special media such as photos, videos, music, or maps, or based on special needs such as shopping, travel, or healthcare. Visualization tools are just turning the corner from research ideas to commercial successes with examples such as the SmartMoney's Map of the Market and improved mapping software.[1]

Consultation tools start with e-mail, but there is much more to be said about this topic (see later).

The four tasks that directly support creative activity are thinking by free association, exploring, composing, and reviewing. Then, once your creative work has been refined, you may want to disseminate it or donate it for the benefit of others. Having an audience or a client in mind helps shape creative projects.

Creativity usually entails an iterative process in which you return to reconsider earlier decisions (figure 10.1). There is rarely a linear path to creative outcomes. You explore possible solutions, and when they don't work out, you backtrack to consider other solutions. Sometimes your breakthrough will come when you go back and redefine the problem itself. Other times you may make progress by visiting your peers and mentors to discuss your directions. Software tools can be helpful.

Searching libraries, the Web, or other resources accelerates your collecting information on previous work. You may also need to go searching in order to find consultants or to decide on candidate communities for disseminating results. Visualizing objects and processes is a task that could appear as part of any of the four activities. Drawing mental or concept maps of your current knowledge helps you organize your knowledge, see relationships, and maybe spot what is missing. Leonardo drew little sketches on many of his pages, integrating his text and images.

Once you've identified a problem and are working on solutions, there are at least four tasks that come up in many discussions of creativity. The most common task is thinking by free association, sometimes called brainstorming. Another popular term, coined by Edward de Bono, is *lateral thinking,* which he defines as "exploring multiple possibilities and approaches instead of pursuing a single approach."

Inspirationalists will appreciate tools to support free association that helps them to break free from their current mindset. Some software tools have attempted to do this by providing related concepts textually, as in the IdeaFisher that provides capabilities that go beyond the usual thesaurus.[2] The developer believes that "creative thinking is an associational process," and he offers a tool to show words that are related to an initial thought by many different cross-referencing paths. Users of IdeaFisher

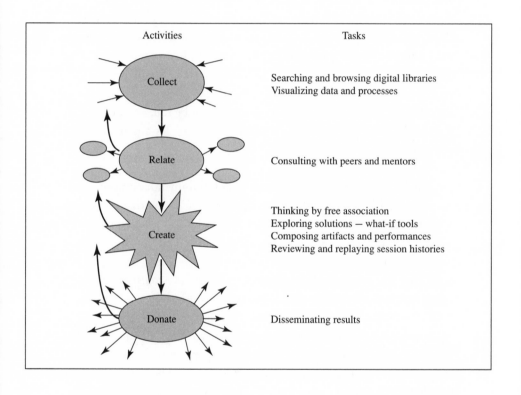

Activities   Tasks

Collect   Searching and browsing digital libraries
Visualizing data and processes

Relate   Consulting with peers and mentors

Create   Thinking by free association
Exploring solutions — what-if tools
Composing artifacts and performances
Reviewing and replaying session histories

Donate   Disseminating results

10.1   Primary relationships of four activities and eight tasks.

offer spirited testimonials to the enjoyable, useful, and sometimes surprising results, for example, "an idea thesaurus of associations on steroids. This program will make you a more effective and persuasive communicator."

Another important task during creative exploration is to run thought experiments about the implications of decisions. Software tools to support this task have matured nicely in many fields. For example, spreadsheets allow you to explore the implications of changes to a business plan, a school budget, or a population growth estimate. More sophisticated tools for simulations have been popping up in every field. These let you set the initial values, try alternative scenarios, and watch what happens. Simulation models have been applied to models of world economies, growth of forests, and collisions of stars. Simulations open your mind to possibilities, allow you to explore safely, and enable you to see complex relationships. Simulations can even be fun, and popular, such as the game-like SimCity, which teaches deep lessons about urban planning:[3]

> SimCity is the first of a new type of entertainment/education software, called system simulations. We provide you with a set of rules and tools that describe, create, and control a real or imaginary system. In the case of SimCity, the system is a city.

SimCity users get to build highways, power plants, stores, and other urban features, but if they don't have enough infrastructure in place, then congestion, pollution, and other problems emerge.

The third set of creativity supports are what I call composition tools. They include the ubiquitous word processor for composing documents and elaborate music editing programs to write symphonies or rock tunes. Graphical composition tools show the enormous power of software to enable more people to be more creative. Slide presentations are now widely used, even by elementary school children, and photo-editing tools have enabled many people to crop, retouch, enhance, and combine their photos.

A compelling composition tool is Dramatica Pro for writing complex movie scripts.[4] It is built on a remarkable theory of story telling and character development that guides you in telling and refining your story.

The successive versions and add-ons to this tool show how software can provide remarkable support for creative productions.

One of the features to be added to many software tools, history keeping, is the capacity to record your activities, review them, and save them for future use. This list lets you return to previous steps, much like the Back button on a World Wide Web browser. But the history-keeping tools I have in mind will also let you edit and replay your history so that you can store frequent patterns of use. You can also send your history list to peers or mentors to ask for help. There is growing evidence that such tools help users and learners in many ways.

Finally, when you have created something you like, you need to disseminate it. Some people will be happy just to send an e-mail to a few friends, but more ambitious possibilities are attractive. During your searching tasks, as you collected information for your work, you encountered the Web sites and the work of many people. So now it might be useful to be able to send your accomplishment to all the people whose work was influential to you. Filters to capture all the e-mail addresses of those people could help you disseminate your work to people who might be interested. An even more ambitious idea would be to send e-mail announcements to all the people who had visited the same Web sites that you visited. The danger of spam (unwanted e-mail) grows quickly, so ways to enable users to specify their interests and willingness to receive unsolicited e-mail must be part of such designs. A more gentle approach is to install your work on a Web page and add entries to indexes that others can explore. Then they can decide about downloading your creative contribution.

I repeat that these eight tasks are not a perfect and complete set. Still, they may be helpful in locating software tools to help you and in identifying candidates for new tools.

CONSULTATION THROUGH NEGOTIATED EXPECTATIONS

Consultation tools deserve special attention because they are such a vital component of creativity support. The problem is that asking for help is difficult for many people; they are uncomfortable about exposing their vulnerabilities. They are afraid to appear uninformed or stupid, and often are not sure what to ask for. They also fear that their time will be wasted with useless responses or that they will be pressured to alter their direction. The biggest source of resistance may come from the fear that their ideas will be stolen. Many inventors have stories to tell of such thefts and had to use the legal system to prove their right to a patent.

In the context of creativity, these resistances take on a special character because of the mystique and uncertain nature of creative explorations. In this regard, Leonardo was like many people. He kept much of his work to himself, writing in a mirror format to hide his insights and conjectures. On the side of openness, he did consult with certain confidants, was very public with his art, and was a lively celebrity, but many of his scientific explorations were quite private.

Leonardo's rivalry with Michelangelo demonstrates how potential colleagues can become rivals. The younger Michelangelo may have learned from Leonardo's work but regularly made disparaging remarks that hurt Leonardo. Michael White writes, "Michelangelo viewed Leonardo as a dilettante, a man who wasted his talents on dreams and gimmicks, someone who could never hold down a commission and who did not deserve the great talent that God had, in his mysterious way, bestowed on him." But Leonardo may have aroused anger by his praise of painting and attack on sculpture—Michelangelo's strength—as a lesser art. Professional rivalries are common to this day, so the reluctance to consult with others is understandable.

My students also resist asking for help, but I insist on collaboration, because I regularly see the improved quality that comes from multiple perspectives. The educational strategies in chapter 6 are built around collaboration and consultation, so that students can learn how to work with others.

The benefits from discussing your work and asking for help seem clear. You may learn more about the domain of application and find out who are the key players in the field. This may lead to your hearing about related work and getting suggestions for improvements. I am constantly amazed by the rich set of perspectives I learn about when talking with others (Okada and Simon 1997). Innovation may seem like a narrow channel with only one way forward; it is a vast ocean with many destinations. A colleague added playfully that there are many ways to slice a tomato.

In addition to providing needed information and fresh ways of thinking, consultants can be helpful because the process of explaining your problem to another person often leads you to a solution. And equally important, if you are struggling with a problem, you may get encouragement to go on, empathy for your struggle, and confirmation of the difficulties you are facing. In short, consultants can give you valuable information and helpful emotional support.

A more complete analysis would include distinctions between early, middle, and late stages of creativity. At the early stages of a project your knowledge is narrow and your needs are broad. You need to learn the domain and field, see previous work, and help focus your attention. During middle stages you have identified some problems and are trying alternative solutions. Thomas Edison methodically tried four thousand materials for light bulbs until he settled on a carbon filament. You may need help choosing candidate solutions wisely. During the late stages you are trying to refine your solution, measure the benefits, and make it work under a variety of conditions. Diverse perspectives may again be helpful.

So the question is how to increase the chance of getting useful help while decreasing the risk of unhappy outcomes? Guarantees are impossible. However, a successful strategy might start with taking small steps onto the beach and proceeding only if the water feels safe and comfortable. You may prefer to approach someone you already know and trust. To avoid misinterpretation, clear messages about who you are and what you want are the way to start. Then if you get a sympathetic response, you might explain why you think your consultant or confidant could be helpful. What does he or she know that you want to know? If you continue to find common ground and a pleasant exchange, then you can make a specific request for help with the time period in which it would be useful. Finally, you get to the delicate matter of offering payment, shared honor, or acknowledgement for help. The following summarizes the components of negotiated expectations for seeking help during creative endeavors:

I clearly identify and refine through dialogue

> *who* I am
> *what* I want to do

and declare my understanding of

> *why* I think you can help
> *how* you can help (specific request with time period)
> *how much* is in it for you (payment, shared honor, appreciation)

You may get a clear yes or no, but often you will have to continue the dialogue to establish clearer understandings and gain higher levels of trust. These negotiated expectations can take some time because of the great uncertainties of the outcomes. Scientists will discuss collaborations and might offer co-authorship or merely an acknowledgment. Physicians or lawyers will consult with one another on an informal basis, but if the consultant wants to get paid, then malpractice and fiduciary responsibility are part of the deal. Negotiating skills are for everyone.

I recall a poorly conceived request from a graduate student in New Zealand by e-mail. He wrote, "Dear Dr. Shneiderman, Attached is my doctoral dissertation proposal. Please let me know if you have any comments." He might have gotten me to respond if he had been clearer about his situation and request. A more appealing note might have said, "Dear Dr. Shneiderman, I am a graduate student working with Prof. Hammond, who knows you. Your work has influenced me and I am extending your interface ideas. The attached proposal (two pages) will be presented in two weeks, so there is time for me to refine it based on your comments. If you are interested, Prof. Hammond says you can be on my dissertation committee and we will fly you to New Zealand for the defence. I would greatly appreciate any help you can offer."

Learning to negotiate expectations could make you more successful in getting assistance in your creative endeavors. You can apply the principles of clear statements to improving your face-to-face encounters, phone calls, or e-mail notes; however, technology improvements might help.

Three directions for new software tools seem possible. First is the enhancement of existing tools (e.g., Microsoft NetMeeting) to enable distant discussants to show each other documents, slide shows, or demonstrations and to jointly control equipment. Improved tools could facilitate richer consultations. Corporate teams could make better business decisions, and orbiting astronauts could get help from developers of scientific payloads. Architects and clients could consult over three-dimensional models and make changes as everyone watches. Faster connections and higher resolution video will have some effect, but complete record keeping (with effective summaries plus bookmarks) and rapid voting will also be useful.

Second, short-term miniconsultations may become more common because of software facilitation. Getting a second opinion from a physician or a lawyer requires substantial time and expense because the legal responsibility or exposure means that professionals must charge full fees. But Web-based miniconsultations might emerge to allow quick second opinions, simple question asking, or guru-for-hire services that reduce expenses by using specialized firms that provide volume services. Of course, there is a need for quality control, but allowing more people to have access to professional consultations at lower cost can produce many benefits. While established professionals will fear their loss of status or income, they may benefit because of referrals and increased interest in professional services.

Third, massive mega-consultations with hundreds or thousands of people participating in government or corporate decisions seem possible. Getting advice can become a massive enterprise, so participatory design methods must be scaled up to support public comments on legislation or customer comments on new products. Mega-consultations have been tried within companies, such as IBM's WorldJam in May 2001, which engaged more than 50,000 employees in brainstorming over corporate directions. Public forums over the Internet to guide government agencies are developing nicely (see chapter 9).

AN ARCHITECTURAL SCENARIO

These ideas for supporting evolutionary creative work are complex. To see how they might be applied, let's imagine how an architect might be engaged in the four activities and the eight tasks. The following scenario has some elements of wishful thinking, but it shows how the activities (collect, relate, create, and donate) and tasks (search, visualize, consult, think, explore, compose, review, and disseminate) might be supported through technology. This scenario makes the revolutionary assumption that creativity support tools would enable an architect to have a broader range of decision-making power, ranging from the initial setting of requirements to the supervision of construction. This reverses the fragmented approach of contemporary architectural practice by restoring control and responsibility to one individual. Architects telling construction companies what to do is the radical idea. However, this is possible only if powerful consultation tools are available to coordinate and supervise tasks that are often delegated to many others.

Imagine an architect, Susan, who is chosen to design a hotel at a national park because of her reputation for flexible designs. For this project she wants to break away from the uniform layout of hotel rooms at many resorts, in which there is a long row of fixed-size identical rooms. She wants to allow flexible modules that can accommodate couples or be reconfigured for families and groups of up to twelve.

To get ideas, she searches an architectural library for digital exemplars of thousands of hotels from around the world, such as Swiss chalets, Austrian lodges, or Rocky Mountain log cabins (collect and search). She selects by size and flips through three hundred possibilities to open up her mind to a broad range of roof designs and sidings (collect and think). She visualizes the data in a two-dimensional scattergram that shows heating requirements, heat loss, and energy consumption patterns for these three hundred possibilities. She uses the interface controls to find strategies that are energy efficient and low-cost (collect and visualize).

Susan chooses a log cabin design and pays the creator a fee, then wrestles with the problems of adding more windows, movable modules, and solar heating panels. Her composition tools allow her to manipulate the underlying architectural model so she can resize the building to accommodate the required number of rooms (create and explore). After choosing a cedar shingle roof and redwood siding, she superimposes the images on the backgrounds of two potential sites: on the hillside and at the base (create and compose). She's ready for her first design review.

The park managers and concessionaires who will run the hotel prefer the hillside site because of the wonderful views. After consulting electronically with park commissioners and travel industry advisers, they accept the log cabin style because it fits local and tourist tastes (relate and consult). A videoconference directly with the client using three-dimensional displays leads to immediate decisions about the placement of the reception desk, a commons area with a fireplace, and a gift shop (relate and consult).

The distinctive plan for flexible modules to accommodate couples, families, and groups of hikers meets resistance, but Susan perseveres by showing how her revised business plan will yield higher revenues through higher occupancy rates. She has added further flexibility, allowing those who wish to cook their own meals to share a communal kitchen/dining area and offering fine dining and maid service for those who want a more pampered experience.

The steep incline of the hillside site presents a formidable challenge, but after an all-night session playing with the engineering models, Susan finds an innovative structural design that costs only 8 percent more than a base site design (create and explore, compose). Susan reflects on how the traditional fragmented approach would have killed the flexible modules idea because independent contractors would not risk such novelty.

Susan collaborates with specialists who confer over her plans for the electric wiring, plumbing, phones, and Internet connections. The same groupware gets her rapid advice but preserves her control over wall decorations, flooring, and furniture styles (create and explore, compose). She does in hours what would have taken weeks when these tasks would have been sent out to consultants. After a virtual walk through, the client requests larger windows, which is handled by reviewing the design history and increasing the window sizes (relate and review). This change causes rerouting of wiring and stronger structural supports, but Susan's flexible design is preserved (create and compose).

At this point, consultations begin with potential builders (relate and consult). Rather than having the park managers do this job, Susan gets them to accept the bold proposal that she handle it herself. Her knowledge of the design and her fluency with the technology make it possible. Susan meets resistance from the builders, but they eventually submit their capabilities report and then their detailed bids electronically to her. She generates bill-of-materials lists for suppliers and a construction schedule for discussion by all parties (relate, create, and compose).

Susan runs into more trouble as she supervises construction. A slow-working subcontractor has attempted to make changes to her design to reduce construction costs, but Susan catches these problems quickly using her process model that checks against digital progress indicators at the construction site. She replaces this subcontractor, and the construction proceeds smoothly during a sunny spring season.

The opening on Memorial Day happens on schedule. There is already a waiting list of couples, families, and groups who are attracted to the beautiful setting and can get just what they want because of the flexible accommodations.

Susan registers her design with the architectural society's digital library and sends a description to interest managers of similar parks around the world (donate and disseminate). Her flexible approach and technology-rich management processes are copied by other architects, for which she collects a fee, and she receives a resort industry award for architectural innovation. Susan appreciates how the software tools enabled creative designs and new business processes but has a list of upgrades she wants for the software before her next project.

THE SKEPTIC'S CORNER

It seems necessary to address the hubris or arrogance of proposing technology to aid human creativity. A critic might say that creativity is inherently human and no computer could or should be brought into the process. But technology has always been part of the creative process, whether in Leonardo's paint and canvas or Pasteur's microscopes and beakers. Supportive technologies can become the potter's wheel and the mandolin of creativity—opening new media of expression and enabling compelling performances. Creative people often benefit from advanced technology to raise their potential and explore new domains.

My expectations are largely positive, but there are many problems, costs, and dangers in anything as ambitious as a framework and tools to support creativity. An obvious concern is that many people may not want to be more creative. Many cultures encourage respect for the past and discourage disruptive innovations. Promoting widespread creativity raises expectations that may change employment patterns, educational systems, and community norms. Introducing computer supports for creativity may produce greater social inequality because it raises the costs for those who wish to participate. Finally, these tools may be used equally by those who have positive and noble goals and by dictators or criminals who seek to dominate, destroy, or plunder.

These fears are appropriate and reasonable cautions must be taken, but support for innovation could lead to positive changes to our world. However, the moral dilemma of technology innovators and users remains troublesome: How can I ensure that the systems I envision or use will bring greater benefits than the negative side effects that I dread and those that I fail to anticipate? Widespread access to effective user interfaces for creativity support could help with major problems such as environmental destruction, overpopulation, poor medical care, oppression, and illiteracy. It could contribute to improvements in agriculture, transportation, housing, communication, and other noble human endeavors.

The path from high expectations to practical action is not easy, but examples of how information technologies helped identify ozone depletion by remote sensing, improved medical diagnosis with computer-aided tomography, and assisted bans on nuclear tests are encouraging. Ensuring more frequent positive outcomes and minimizing negative side effects remain challenges, but a framework that provides for substantial consultation and broad dissemination may help. Participatory design methods and comprehensible, widely disseminated social impact statements may be effective because they promote discussion and expand the range of options for decision makers.

Portrait of a Musician. From license-free "Leonardo da Vinci: Selected Works," Planet Art.

11

> GRANDER GOALS

Unlike machines, human minds can create ideas. We need ideas to guide us to progress, as well as tools to implement them. . . . Computers don't contain "brains" any more than stereos contain musical instruments. . . . Machines only manipulate numbers; people connect them to meaning.
—Arno Penzias, *Ideas and Information* (1989), 179

THE OLD COMPUTING AND THE NEW COMPUTING

Computing in the twenty-first century will be shaped by the pace of the shift from the old computing to the new computing. There is still a need for researchers to build faster machines and write faster algorithms. Large measurable benefits can emerge from the work of the old computing, for which developers can still receive commercial rewards and academic honors. One colleague reported a dramatic example involving a standard mathematical problem: in 1990, it required twelve hours to produce a solution on a personal computer, but by 2000, with improved hardware and software, it required only 6 seconds.

Contributions to speeding up computers will continue to be valuable, as will software research to compress data and reduce costs. However, I believe that attention and funding will shift to the new computing ideas that focus on human needs and to the old computing ideas that most directly support this shift. New computing advances that improve your trust, privacy, and security will increase your capacity and desire to participate in markets and contribute as citizens. As universally usable designs spread, the expanding user community will dissolve the digital divides. These improvements will amplify your ability to form satisfying relationships with family, friends, neighbors, and colleagues, and your capacity to carry out creative endeavors.

The pace at which public interest and investment shifts toward new computing projects will determine how soon improvements appear. Changes in priorities at government agencies and corporate research centers are likely to be controversial. Therefore, consumer activists and public-spirited journalists can be influential in accelerating the movement to a new consensus. Then funders will have greater clarity in choosing among the proposals made by promoters of high-speed compilers versus universal usability or database optimization versus online help. They will confidently choose proposals that will do the most to increase your capacity to collect information rapidly, establish communication effectively, innovate creatively, and disseminate results widely.

The most contentious battles are likely to form around proposals from researchers who want to build exotic computers and applications that have little to do with improving the user experience. I'm especially critical of those members of the artificial intelligence community who propose to build machines that do tasks that humans do rather than empowering humans to do the tasks. For instance, Dr. John McCarthy, one of the pioneers of artificial intelligence, writes, "The ultimate effort is to make computer programs that can solve problems and achieve goals in the world as well as humans."[1] Similar statements have been made over the years by others, but this mimicry game or replacement theory has little value in improving the user experience.

The allure of intelligent machines with humanlike qualities has at least a two-thousand-year history. Homer described Hephaistos, the god of fire and forging, and his golden robots with these images (*Iliad* 18.481):

> He limped out of his workshop. Round their lord came fluttering maids of gold, like living girls: intelligences, voices, power of motion these maids have, and skills learnt from immortals. Now they came rustling to support their lord.

Eighteenth-century clockmakers fabricated robot dolls that played musical instruments, wrote poems, and drew pictures. But these amusements for royalty became little more than museum pieces for the next century. The most advanced technologies are often applied by dreamers who wish to fabricate a humanlike robot. During the twentieth century computing technology became the stimulus for efforts to build humanlike robots that cleaned your home or rolled through office corridors delivering mail. Industrial robots were also developed by initially mimicking human form, for example, with five-fingered hands that could only lift thirty pounds and wrists that rotated only halfway around.

Meanwhile nonanthropomorphic (not based on human form) technologies that were not influenced by the humanlike robotic notions did spread. Washing machines became success stories in most modern homes, and flexible-manufacturing systems that could rapidly place five-hundred-pound car engines precisely on their frames were deployed worldwide. Paying attention to human needs at home or in factories leads to more rapid progress than trying to make an artificial humanlike robot. Of course, humanlike robots are useful for some things, such as Disney's audio-animatronic entertainment and more realistic automobile crash-test dummies. There is also important work being done on prosthetic devices where the goal is to replace lost human capabilities.

Journalists of the 1940s nurtured the notion of computers as "awesome thinking machines" with headlines such as "Electronic Brain Thinks Faster Than Einstein" and "Magic Brain Spurs Science and Technology." As Dianne Martin (1993) details in her scholarly analysis, the public attitude was profoundly shaped by these reports. Surveys in 1963 and 1983 found strong public support for the belief that computers "can think like a human being thinks" and "sort of make you feel that machines can be smarter than people." She conjectures that the "awesome thinking machine" myth may have in fact retarded public acceptance of computers in the work environment at the same time as it raised unrealistic expectations for easy solutions to difficult social problems.

Contemporary variations focus on the possibility of building machines that are emotionally aware and even affectively responsive. These new visions suggest that computers could recognize your growing anxiety or frustration and offer calming or reassuring responses. Some researchers avoid the suggestion that the machine experiences emotion, but others rise to the challenge of building a fully emotional and conscious robot. Steven Spielberg's 2001 movie, *A. I.* (for artificial intelligence), showed the product of a project to build a child that loves. This Hollywood portrayal encourages the fantasy, but the story's unhappy outcome will discourage at least a few proponents.

Even serious scientists are prone to considering artificial consciousness as a good, useful, and attainable research goal. An article (Buttazzo 2001) in the highly respected *IEEE Computer* magazine claims, "Many people now believe that artificial consciousness is possible and that, in the future, it will emerge in complex computing machines." That enthusiasm might be traced to Ray Kurzweil's *The Age of Spiritual Machines: When Computers Exceed Human Intelligence* (1999).

Kurzweil's extrapolation of Moore's law leads him to conclude that by 2020 computers will have more circuits and connections than there are neurons in the human brain. He allows for some additional overhead to organize machine thinking but confidently predicts, "Before the next century is over, human beings will no longer be the most intelligent or capable type of entity on the planet. . . . Machines, derived from human thinking and surpassing humans in their capacity for experience, will claim to be conscious, and thus to be spiritual." This statement strikes me as preposterous; I think machine claims to spirituality are about as important to scientific progress as a drunkard's claim to be a god.

The questionable path that Kurzweil pursues is to use the metrics of the old computing while failing to grasp the centrality of human experience in the new computing. He considers the calculations per second per cubic centimeter of brains and computers but doesn't count the inspirations per hour from a mentor or the trajectory of trust while talking to your doctor. His focus is on machines replacing or competing with people rather than on tools to help people achieve more of what they want. While computers will undoubtedly grow more powerful, human needs are central to me. For Kurzweil, values and emotions are merely "unavoidable by-products of the levels of abstraction that we deal with as human beings." Trust and empathy don't make it into his index. At least Hollywood deals with emotional issues among family, friends, and colleagues, but human relationships hardly appear in Kurzweil's world.

Kurzweil will have his disciples, but I hope most people will recognize the vitality of Mumford's description of technology goals, as "serving human needs." If researchers and developers create innovations that empower rather than replace people, they will be more likely to accelerate productive technology evolution. Tools that support doctors in making better diagnoses than any doctor has ever done are more likely to succeed than systems that replace doctors. Future physician support tools will enable genetic analyses to be applied to diagnostic decisions and simulated treatments to be tested against human physiological models. Educational discussion groups and e-mail exchanges with professors are more likely to spread than intelligent tutoring systems that replace teachers. Future educational tools will enable students to collaborate in performing complex scientific experiments or comparing theories of child development. Electronic markets like eBay will flourish as they become more enticing and effective, while intelligent agents that do your shopping for you are likely to remain marginal. Future shopping experiences will enable customers to get trustworthy comparison data and to organize cooperative buying groups.

Successful technology developments will come from those who recognize the importance of tools and social systems that support human goals, control, and responsibility. Users want the sense of mastery and accomplishment that comes from using a tool to accomplish their goals. At time passes, the range of automation will expand, but only if users can understand and control what is happening. Turning the key in your automobile or stepping on the accelerator invokes many complex computer calculations, but the results are comprehensible, predictable, and controllable. Similarly, even sophisticated technologies such as cruise control give you more precise control over your speed with less effort. Of course, there are exceptional times when rapid automated action to prevent destruction is needed, such as in antilock brakes or air bags.

The role of social systems is also powerful in determining technology acceptance and adoption. The strong human desire for relationships is what drives the success of e-mail, chat rooms, and online communities. These have nothing to do with the replacement theory and everything to do with supporting human social relationships. A second role of social systems is to cope with technology breakdowns or times when users need to go beyond the design limits of machines. Human relationships and trust are needed to resolve such problems, cope with the failures, and seek appropriate improvements.

The goal of making computers do what humans do, the replacement theory, also seems rather modest. Imagine proposing to make a bulldozer that lifts as much as the strongest human or a printer that writes as fast as the best human scribe. Truly beneficial technologies should amplify human capacity by a factor of a hundred or more. But speed is only one beneficial attribute; quality is another. Speed is easy to measure, but quality improvement is more difficult to assess.

Quality measures come from subjective reviews by peers, from dialogue that furthers understanding, and from participatory processes that establish community norms. This is difficult to accept for many workers of the old computing who want objective and even automatable metrics. Advocates of the new computing often seek subjective measures and apply ethnographic approaches to conduct evaluations. They observe and interview users while they are doing their work or enjoying their entertainment. The results are not numbers but understanding, not percentages but insights.

I hope I have convinced you of the importance of empowering, not replacing, users in the design of new technology. Time and again the success stories have been written by those who understood this principle: graphical user interfaces, word processors, multimedia, e-mail, the World Wide Web, instant messaging, online communities, information visualization, and e-business. However, there are still loftier goals that will guide us for the future, so we need to go around to a higher point on the helix of human needs.

Building a computer that empowers users is a noble goal, even for the four basic applications in this book: e-learning, e-business, e-healthcare, and e-government. These have immediate payoffs for students, commercial potential for merchants and customers, practical benefits for physicians and patients, and improvements for citizens. Of course, there are other important applications that could be covered, such as appropriate housing, safe transportation, and socially constructive entertainment and sports. Then there are higher human values that we must aspire to serve. We can pursue environmental quality and quality of life. We can strive to resolve conflicts and promote peace.

Responding to these grand concerns and enduring values may seem to be beyond the scope of users and technology developers, but I believe that you can attain them by focusing on specific and measurable goals such as the following:

> Increase life expectancy.
> Control population growth.
> Reduce homelessness.
> Reduce illiteracy worldwide.
> Reduce automobile accident deaths.
> Increase air quality in major cities.
> Reduce the threat of war.

A dramatic example of the power of individual activism to bring peaceful change comes from the story of Jody Williams (Wildmoon 1997). This forty-seven-year-old Vermont activist used e-mail to create an international movement against the use of land mines. Her work led to an international treaty banning land mines and earned her the Nobel peace prize in 1997. Mary Wareham, who was a coordinator of the U.S. Campaign to Ban Landmines, described the efficacy of e-mail: "It's fast, convenient, easy to use, and it's cheap . . . and it's effective. (E-mail) played a very important role. . . . We created the momentum for this political process."

Other examples of how technology can promote peaceful outcomes are becoming clearer. Some commentators credit the hot line, an electronic communication network between the Washington's White House and Moscow's Kremlin, for reducing tensions during two decades of the cold war.

In my personal experience, I had the satisfaction of working with the devoted staff of the International Atomic Energy Agency in Vienna to improve tools and techniques for inspectors who monitor the Nuclear Non-Proliferation Treaty. Teams of inspectors visit nuclear power and research facilities in 120 countries to verify adherence to the treaty. Often work must be stopped during inspections, so rapid, error-free performance in no more than one to three days is essential. However, the inspectors struggle to use the ninety different detectors and other instruments, which have been developed by different suppliers from many ountries. Getting help is time-consuming and disruptive, so the inspectors are under great pressure. My job was to help them develop standards for terminology, units, and displays, with simplified operating procedures. This experience also introduced me to the important work of computer scientists to make possible the Nuclear Test Ban Treaty by developing seismic monitoring software to detect nuclear bomb tests. It was satisfying to see how computing could be applied to promoting peace.

In many other creative ways, information and computer technology can make a direct impact, for example, by educational applications in literacy training or by computer control of automobile engines to reduce pollution. In other situations, the linkage with improved human-computer interaction may be less clear initially. In fact, some goals may be difficult to attain by only redesigning computer technology, but the example of increased public interest and engaged computing professionals may prove to be an inspiration to others. Therefore, even though we may not know the path, a clear statement of the destination will help attain those goals and inspire broad participation by others.

The old computing was often seen as a source of frustration, a threat to privacy, a tool of oppression, or a source of destruction. These aspects generated a wide range of technology critics who pointed out failures and sought to limit the growth of information and computing technologies. These critics could not imagine how constructive criticisms could produce more positive outcomes. They did not anticipate the emergence of new computing, which promotes more positive human values.

And always we must remember that low or no teehnology may also be a constructive path. The absence of cell phones enhances our theaters, and the freedom from automobiles enriches our experience of hiking in the woods. Calmness when we are alone, intimacy with friends and family, and close contact with colleagues and neighbors might also be listed in a technology-free activities and relationship table. We should question every application of technology to ensure that its benefits outweigh its costs.

THE NEXT LEONARDO

Leonardo has served well as an inspirational muse for the new computing. I hope you enjoyed the excursions to imagine how Leonardo's life might have lessons for the new computing. So let's conclude by taking on the question of how the new computing might encourage the emergence of the next Leonardo.

How will we recognize the next Leonardo? Do we look for illegitimate children from mountain villages or vegan artists who sketch human anatomy? I think the essence of Leonardo was his capacity to think with originality about the grand questions of his time and to innovate widely. These skills emerged in Leonardo from his ceaseless questioning about the world followed by his careful observation and focused experimentation. He pushed aside fifteen centuries of devotion to Aristotle's logic and began the modern traditions of scientific inquiry by proposing hypotheses and seeking empirical evidence.

A modern Leonardo, let's call him or her Leonardo II, would be asking questions in a wide range of disciplines and be testing ideas with pilot projects and rapid prototypes. Leonardo II would also be an effective promoter of ideas who spoke to the media and put on compelling performances. Leonardo II might be a genomic researcher like Craig Ventner, who masters television commentary with the authoritative tone of naturalist David Attenborough and the political savvy of Gandhi. A modern Leonardo might also be a musical performer like Madonna, who has the eye for color and form of Georgia O'Keeffe and the computing insights of Bill Gates. In our specialist age, combinations of multiple skills surprise us, but that is what made Leonardo so amazing and creative.

Among Leonardo II's skills would be deft use of computers for collecting information and relating to peers by instant messaging or e-mail. He or she would master a variety of creativity support tools and would make new ones regularly as gifts for friends and patrons. He or she

242 > Chapter 11

would carry portable devices for quick notes and sketches, plus larger
tablets for lengthier compositions and bigger drawings. Leonardo II
would install projectors in a studio to examine large images with guests
and create wall-sized tableaus. She or he would record thoughts on Web
sites, disseminating some ideas widely but keeping most files password-
protected for future publication.

It's easy to get carried away by this fantasy, but beyond the daily life
choices, Leonardo II would be committed to at least two larger
principles:

Technical excellence must be in harmony with user needs.

Great works of art and science are for everyone.

Devotees of artificial intelligence who pursue the replacement scenarios would undoubtedly tell a different story. They would probably
predict that Leonardo II would be a computer, based on neural nets,
genetic algorithms, and semantic understanding programs. Their version
of Leonardo II would have full natural language speech interaction, emotional responsiveness, and creativity accelerator hardware chips. I suppose
I should smile at this notion, but someone might start to build it. That
would be fine if they did this on their own time, but if public funds
were being spent, then I'd rather see the money devoted to teaching children about Leonardo.

THE SKEPTIC'S CORNER

We must always remember the danger that malicious individuals, organizations, and nations may apply technology for destructive purposes. As
much as we wish that innovations would be put to work to benefit people, there is always a threat of more troubling scenarios. It would be
tragic if the main outcome of advanced technology were to empower dic-

tators, criminals, and terrorists. We can't prevent them from gaining access to widespread technologies, so we must anticipate that some nasty applications will emerge. As users, we must be alert to the dangers in our lives, make our best effort to stop malevolent applications, and support social processes that limit their growth. As developers, we can choose our applications carefully to show positive contributions, open our thinking to anticipate unintended negative side effects, and expose our work to participatory design reviews. Open discussion, open code, and open ideas are among the greatest benefits of information and communication technologies.

A skeptic might fear that there will never be another Leonardo and that technology progress is more harmful than helpful. Such pessimistic views go too far because it seems clear that human beings are enriched by their tools and social structures. Throughout history those who employed effective technologies and social structures flourished. We benefit daily from medical care, safe transportation, and improved education. To achieve the positive and lofty goals proposed in this chapter, public awareness and the consciousness of designers must be raised. An open international debate that stimulated discussions among individuals and within organizations could help promote more positive outcomes. Those who believe that they can shape the future will shape the future.

Chapter 1

1. The attraction of Leonardo's sculpture was so compelling that exactly five hundred years after the model was destroyed—on September 10, 1999—an American business-man, Charles Dent, fulfilled Leonardo's dream by producing the bronze horse and installing it in Milan.

2. National Gallery of Art <http://www.nga.gov>, which is home to Leonardo's paint-ing of Ginevra de' Benci <http://www.nga.gov/cgi-bin/pinfo?Object= 50442+0+none>.

3. America Online (AOL) has more than 30 million users because the company has made a strong commitment to user control and ease of use. Its discussion-oriented tool, ICQ (at <http://www.icq.com/>), is widely used. ICQ is described as "a user-friendly Internet program that notifies you which of your friends and associates are online and enables you to contact them." The ICQ service claims more than 105 million users worldwide and is growing rapidly because it serves human needs and is well designed to support those needs.

Chapter 2

1. See Peter Neumann's Risks Forum, <ftp://ftp.sri.com/risks>. See also <http://www.csl.sri.com/users/neumann/neumann.html>.

2. "Sea of Lies," March 20, 1988, <http://www.geocities.com/CapitolHill/5260/vince.html>. See also <http://catless.ncl.ac.uk/Risks/8.74.html#subj1>.

3. A creative idea for end user debugging when crashes occur is described in an appro-priately angry article by Henry Lieberman and Christopher Fry, "Will Software Ever Work?" *Communications of the ACM* 44 (March 2001): 122–124.

4. See <http://www.geoman.com/Vitruvius.html>.

5. Netscape Quality Feedback System for Netscape Communicator 4.5, <http://home.netscape.com/communicator/navigator/v4.5/qfs1.html>.

6. Webby Awards, <http://www.webbyawards.com/main/>.

Chapter 3

1. From CyberAtlas, Geographics, <http://cyberatlas.internet.com>.

2. United Nations Development Programme, <http://www.undp.org/>; United Nations Information Technology Service, <http://www.unites.org/>; Partnerships Online: Creating Online Communities for Neighbourhoods and Networks, <http://www.partnerships.org.uk/>; Volunteers in Technical Assistance, <http://www.vita.org/>.

3. Digital Divide Network, <http://www.digitaldividenetwork.org/>, developed by the Benton Foundation, <http://www.benton.org/>.

4. See <http://www.conversa.com/>.

5. See Neil Scott's Archimedes Project at Stanford University, <http://archimedes.stanford.edu/>.

6. The U.S. National Cancer Institute is at <http://www.nci.nih.gov/>, and the cancer information is at <http://www.nci.nih.gov/cancerinfo/index.html>.

7. NASA's regular page is at <http://www.nasa.gov/>, and the page for kids is at <http://www.nasa.gov/kids.html>.

8. Altavista, <http://world.altavista.com/>, provides the Systran services, <http://www.systransoft.com/>. The Seattle Community Network offers a wonderful guide to resources for translating its content to many languages, <http://www.scn.org/spanish.html>.

9. There are several resources for universal design. *Professional groups:* The ACM SIGCHI (Special Interest Group on Computer-Human Interaction), <http://www.acm.org/sigchi/>, focuses on design of useful, usable, and universal user interfaces. SIGCHI promotes diversity with its outreach efforts to seniors, kids, teachers, and international groups, and it sponsors the Conferences on Universal Usability, <http://www1.acm.org/sigs/sigchi/cuu/>. The ACM's SIGCAPH (Special Interest Group on Computers and the Physically Handicapped), <http://www.acm.org/sigcaph/>, has long promoted accessibility for disabled users, and its ASSETS series of conference proceedings, <http://www1.acm.org/sigs/sigcaph/assets/>, provides useful guidance. The European conferences on User Interfaces for All, <http://ui4all.ics.forth.gr/index.html>, also deal with interface design strategies. The Web Accessibility Initiative, <http://www.w3.org/WAI/>, of the World Wide Web Consortium has a guidelines document with fourteen thoughtful content design items to support disabled users. *Corporate Web sites:* Sun Microsystems, <http://www.sun.com/access/>, offers Java-specific recommendations. Thoughtful Web sites from IBM, <http://www.ibm.com/easy/>, and Microsoft, <http://www.microsoft.com/enable/>, describe processes and designs for supporting diverse users. *University Web sites:* North Carolina State University's Center for Universal Design, <http://www.design.ncsu.edu/cud/>, lists seven key principles, and the University of Wisconsin's TRACE Center, <http://trace.wisc.edu/world/>, offers links to many resources. Another source, <http://universalusability.org/>, has a taxonomy of topics and links, plus information on the Universal Usability Policy template for inclusion on Web sites. Students at the University of Maryland have created a Universal Usability in Practice Web site, <http://www.otal.umd.edu/uupractice/>, with design guidelines.

Chapter 4

1. National Institute of Standards and Technology, <http://www.nist.gov/iusr/>.

2. ACM SIGCHI, <http://www.acm.org/sigchi/>.

3. Human-Computer Interaction Institute, Carnegie-Mellon University, <http://www.hcii.cmu.edu/>; Media Lab, Massachusetts Institute of Technology, <http://www.media.mit.edu/>.

4. Stanford University Program in Human-Computer Interaction, <http://hci.stanford.edu/>; University of Maryland Human-Computer Interaction Lab, <http://www.cs.umd.edu/hcil/>.

Chapter 5

1. An extensive Web site with multiple data sets is available at <http://webuse.umd.edu/>.

2. Maslow (1968). A thoughtful review of Maslow's ideas appears at <http://www.ship.edu/~cgboeree/maslow.html>.

3. eBay, <http://www.ebay.com/>; the Nasdaq Stock Market, <http://www.nasdaq. com/>; Amazon, <http://www.amazon.com/>.

4. U.S. Library of Congress, <http://www.loc.gov/>.

5. New York Stock Exchange, <http://www.nyse.com/>; Fidelity Investments, <http:// www.fidelity.com/>; Smartmoney, <http://smartmoney.com/>; Charles Schwab & Co., <http://www.schwab.com/>.

6. I've been profoundly influenced by Putnam's *Bowling Alone* (2000). His analysis of the decline of "social capital" since 1965 is a brilliant explanation of how and why Americans have reduced their participation in community groups and political activities. He even demonstrates reduced participation in picnics and dinner parties. The explanations and documentation of harm are disturbing.

7. American Memory Project at U.S. Library of Congress, <http://memory.loc.gov/>.

8. PictureQuest, <http://www.picturequest.com/>; Corbis, <http://www.corbis.com/>.

9. Bederson (2001). PhotoMesa is free to download at <http://www.cs.umd.edu/hcil/ photomesa/>.

10. Brooklyn, <http://www.brooklyn.com/>; Prospect Park neighborhoods, <http:// www.prospectpark.org/>; Brooklyn history, <http://www.brooklynhistory.org/>.

11. IBM archives, <http://www-1.ibm.com/ibm/history/>; Intel, <http://www.intel. com/intel/intelis/museum/index.htm>.

12. Kodak, <http://www.kodak.com/>, "Quilt".

Chapter 6

1. Csikszentmihalyi (1996). His book was a wonderful discovery and very influential in my thinking.

2. My appreciation of this topic has emerged over my ten years of teaching in our advanced technology classroom, AT&T Teaching/Learning Theater. We discussed educational philosophies strenuously as a guide to designing and then using this classroom and its successors, <http://www.inform.umd.edu/TT/>. The University of Maryland Teaching Theaters Steering Committee was a lively forum for these discussions. Key researchers were Maryam Alavi, Kent Norman, Jim Greenberg, Glen Ricart, and Ellen Yu Borkowski. My writings on this topic (Shneiderman 1989; 1992; 1998) were accompanied by a team effort to review and document the four collaborative styles that evolved among the 75+ faculty members who taught three hundred courses there (Shneiderman et al. 1995; 1998).

3. WebCT: Helping Educators Transform Education, <http://www.webct.com/>; Blackboard: Bringing Education Online, <http://www.blackboard.com/>.

4. Davidson and Worsham (1992); Millis (1990). Neil Davidson and Barbara Millis were both helpful guides in introducing me to collaborative teaching methods and their enormous value.

5. GroupSystems, <http://www.groupsystems.com/>.

6. The evidence for the benefits of collaboration over competition has built up over the years and is wonderfully presented in Kohn (1986).

7. This distance learning course in fall 1993 had twelve students in front of me and twelve more who watched via satellite TV. The first student team project, completed in six weeks, was the Encyclopedia of Virtual Environments, <http://www.hitl.washington.

edu/scivw/EVE/>. This was followed by the Journal of Virtual Environments, completed at the end of the fifteen-week semester, <http://www.hitl.washington.edu/scivw/JOVE/>. These projects were eagerly taken over by the very helpful cybrarian Toni Emerson at the University of Washington Human Interface Technology Lab, which specializes in virtual reality.

8. Spring semester project, 2001, <http://www.otal.umd.edu/uupractice/>.

9. My students named the project Web site Student Human-Computer Interaction Online Research Experiments—SHORE—to conjure up an image of Maryland's beaches, <http://www.otal.umd.edu/SHORE97/>; also, < . . . /SHORE98/>, < . . . /SHORE99/>, < . . . /SHORE2000/>, and < . . . /SHORE 2001/>.

10. Jacoby et al. (1996). Barbara Jacoby crusaded for community service projects on the College Park campus and pushed this idea vigorously in her book.

Chapter 7

1. Buy Brigade, <http://www.cnet.com/>.

2. i411 Interactive Information Discovery, <http://www.i411.com/>.

3. Better Business Bureau Online, <http://www.bbbonline.com/>.

4. eComplaints, <http://www.ecomplaints.com/>.

5. Priceline, <http://www.priceline.com/>.

6. Personalization, <http://www.personalization.com/>.

7. eBay auctions, <http://www.ebay.com/>, "Feedback Forum".

8. TRUSTe, <http://www.truste.com/>.

9. Electronic Privacy Information Center, <http://www.epic.org/>.

10. SquareTrade, <http://www.squaretrade.com/>.

Chapter 8

1. Merck Manual Home Edition, <http://www.merckhomeedition.com/>.

2. Yahoo! groups, <http://groups.yahoo.com/>.

3. Computer-Based Patient Record Institute and Healthcare Open Systems and Trials, <http://www.cpri-host.org/>.

4. Maryland Board of Physician Quality Assurance, <http://www.bpqa.state.md.us/>.

5. National Institutes of Health, National Human Genome Research Institute, <http://www.nhgri.nih.gov/>.

6. IBM Blue Gene Project, <http://www.research.ibm.com/bluegene/>.

7. U.S. National Library of Medicine, <http://www.nlm.nih.gov/>.

8. The National Institutes of Health provide current information about clinical research studies, <http://clinicaltrials.gov/>.

9. WebMD, <http://webmd.com/>; Dr. Koop, <http://www.drkoop.com/>.

10. CompuMentor, <http://www.compumentor.org/>; TechSoup, <http://www.techsoup.org/>.

11. The Bill and Melinda Gates Foundation, <http://www.gatesfoundation.org/>, set a positive example in their strong commitment to supporting vaccinations, HIV/AIDS treatment, and children's health in many Third World countries.

12. United Nations Information Technology Service, <http://www.unites.org/>.

13. Doctors Without Borders, <http://www.doctorswithoutborders.org/>; Med Help International, <http://medhelp.org/>.

14. This playful fantasy is based on Frank Baum's classic story, *The Wizard of Oz*, which was made into a 1939 film starring Judy Garland as Dorothy. For readers who don't know this story: Toto was Dorothy's dog and the Cowardly Lion (the inspiration for Dr. Lyon) was one of Dorothy's companions during her adventure. Dorothy received a magical pair of Ruby Red Slippers from the Good Witch of the North. The Munchkins were happy characters that Dorothy encountered in the forest. This fanciful scenario originated in my 1998 keynote address in Los Angeles to ACM's SIGCHI Conference on Human Factors in Computing Systems. An electronic demonstration of the medical system created in Macromedia Director by Chris North is at <http://www.cs.umd.edu/hcil/>, "Genex".

15. LifeLines is a research information visualization tool for exploring temporal history data such as is found in patients' records. Facets, such as doctor visits, hospitalizations, lab tests, and medications, are shown in parallel horizontal panels. LifeLines is based on research at the University of Maryland Human-Computer Interaction Lab, <http://www.cs.umd.edu/hcil/lifelines>.

16. Spotfire, <http://www.spotfire.com/>, is a commercial information visualization tool for exploring complex data. It grew out of research at the University of Maryland Human-Computer Interaction Lab. Its main successes have been in pharmaceutical drug discovery and DNA microarray data analysis.

Chapter 9

1. Washington State, <http://access.wa.gov/>.

2. City of Santa Monica, <http://pen.ci.santa-monica.ca.us/cm/>.

3. Seattle Community Network, <http://www.scn.org/>; see also Schuler (1996).

4. U.S. Conference of Mayors, <http://www.usmayors.org/>.

5. For technologists, this is where XML tags come in, to provide a more semantically organized Web. The coordination must still be done by people to ensure effective standard definitions.

6. U.S. Census Bureau, <http://www.census.gov/>.

7. Intuit, Quicken TurboTax, <http://www.quicken.com/taxes/>.

8. Slashdot, <http://slashdot.org/>, is a public discussion forum.

Chapter 10

1. SmartMoney, <http://smartmoney.com/>; Environmental Systems Research Institute, <http://www.esri.com/>.

2. One of my pleasures in preparing this book was to have an hour's phone conversation with Marsh Fisher, the developer of IdeaFisher, <http://www.ideafisher.com/>.

3. SimCity, <http://simcity.ea.com/>.

4. Dramatica, <http://www.dramatica.com/>.

Chapter 11

1. Dr. John McCarthy at <http://www.kurzweilai.net/articles/art0088.html?printable=1>.

REFERENCES

AAHE (American Association for Higher Education). 1987. *Principles for Good Practice in Undergraduate Education.* Washington, D.C.

Access Board. 2000. Electronic and Information Technology. <http://www.access-board. gov/sec508/status.htm>.

Alavi, Maryam. 1994. Computer-Mediated Collaborative Learning: An Empirical Evaluation. *MIS Quarterly* 18 (2): 159–173.

Ausubel, David. 1968. *Educational Psychology: A Cognitive View.* New York: Holt, Rinehart and Winston.

Baecker, R., K. Booth, S. Jovicic, J. McGrenere, and G. Moore. 2000. Reducing the Gap Between What Users Know and What They Need to Know. In *Proceedings of the ACM Conference on Universal Usability,* 17–23. New York: ACM Press.

Bederson, Benjamin. 2001. Quantum Treemaps and Bubblemaps for a Zoomable Image Browser. In *Proceedings of User Interface Software and Technology Symposium 2001.* New York: ACM Press.

Bergman, Eric, ed. 2000. *Information Appliances and Beyond.* San Francisco: Morgan Kaufmann.

Boden, Margaret. 1990. *The Creative Mind: Myths and Mechanisms.* London: Weidenfeld and Nicolson.

Brooks, Frederick, Jr. 1996. The Computer Scientist as Toolsmith II. *Communications of the ACM* 39 (3): 61–68.

Bush, Vannevar. 1945. As We May Think. *Atlantic Monthly* 76 (July): 101–108. Also at <http://www.theatlantic.com/unbound/flashbks/computer/bushf.htm>.

Buttazzo, Giorgio. 2001. Artificial Consciousness: Utopia or Real Possibility? *IEEE Computer* 34 (7): 24–30.

Card, Stuart, Jock Mackinlay, and Ben Shneiderman, eds. 1999. *Readings in Information Visualization: Using Vision to Think.* San Francisco: Morgan Kaufmann.

Carroll, J., and C. Carrithers. 1984. Training Wheels in a User Interface. *Communications of the ACM* 27 (8): 800–806.

Carson, Rachel. 1962. *Silent Spring.* Boston: Houghton Mifflin.

Cave, Charles. 2001. Creativity Web. <http://members.ozemail.com.au/~caveman/ Creative/index2.html>.

Clark, Kenneth. 1939. *Leonardo da Vinci.* Rev. and introduced by Martin Kemp. London: Penguin Books, 1988.

———. 1966. On the Relation Between Leonardo's Science and His Art. In *Leonardo da Vinci: Aspects of the Renaissance Genius,* ed. Morris Philipson. New York: Braziller.

Compaine, Benjamin, ed. 2001. *The Digital Divide: Facing Crisis or Creating a Myth?* Cambridge, Mass.: MIT Press.

Corbis, Inc. 1997. *Leonardo da Vinci.* CD-ROM.

Couger, J. D. 1996. *Creativity and Innovation in Information Systems Organizations.* Danvers, Mass.: Boyd and Fraser.

Covey, S. R., A. R. Merrill, and R. R. Merrill. 1994. *First Things First: To Live, to Love, to Learn, to Leave a Legacy.* New York: Simon and Schuster.

Csikszentmihalyi, Mihaly. 1996. *Creativity: Flow and the Psychology of Discovery and Invention.* New York: HarperPerennial.

CSTB (Computer Science and Telecommunications Board). National Research Council. 1997. *More Than Screen Deep: Toward Every-Citizen Interfaces to the Nation's Information Infrastructure.* Washington, D.C.: National Academy Press.

Davidson, Neil, and Toni Worsham. 1992. *Enhancing Thinking Through Cooperative Learning.* New York: Teachers College Press.

de Bono, Edward. 1973. *Lateral Thinking: Creativity Step by Step.* New York: HarperCollins.

Dearing, Ron, chair. 1997. National Committee of Inquiry into Higher Education. Report. <http://www.leeds.ac.uk/educol/ncihe/>.

Dewey, John. 1916. *Democracy and Education.* New York: Macmillan.

Druin, Allison. 1999. Cooperative Inquiry: Developing New Technologies for Children with Children. In *Proceedings of CHI 99, Conference on Human Factors in Computing Systems,* 592–599. New York: ACM Press.

Druin, Allison, B. Bederson, J. P. Hourcade, L. Sherman, G. Revelle, M. Platner, and S. Weng. 2001. Designing a Digital Library for Young Children: An Intergenerational Partnership. In *Proceedings of ACM/IEEE Joint Conference on Digital Libraries,* 398–405. New York: ACM Press.

Druin, Allison, and James Hendler, eds. 2000. *Robots for Kids: Exploring New Technologies for Learning.* San Francisco: Morgan Kaufmann.

Druin, Allison, J. Montemayor, J. Hendler, B. McAlister, A. Boltman, E. Fiterman, A. Plaisant, A. Kruskal, H. Olsen, I. Revett, T. Plaisant-Schwenn, L. Sumida, and R. Wagner. 1999. Designing PETS: A Personal Electronic Teller of Stories. In *Proceedings of CHI 99,* 326–329. New York: ACM Press.

Fogg, B. J., and H. Tseng. 1999. The Elements of Computer Credibility. In *Proceedings of CHI 99,* 80–87. New York: ACM Press.

Fountain, Jane. 2001. *Building the Virtual State: Information Technology and Institutional Change.* Washington, D.C.: Brookings Institution.

Frere, Jean Claude. 1995. *Leonardo: Painter, Inventor, Visionary, Mathematician, Philosopher, Engineer.* Paris: Terrail.

Freud, Sigmund. 1910. *Leonardo da Vinci and a Memory of His Childhood.* Ed. James Strachey, trans. Alan Tyson. New York: Norton, 1990.

Friedman, Thomas. 2000. *The Lexus and the Olive Tree: Understanding Globalization.* New York: Farrar, Straus, Giroux.

Fry, Christopher, and Henry Lieberman. 1995. Programming as Driving: Unsafe at Any Speed? In *Proceedings of CHI 95, Demonstrations,* 3–4. New York: ACM Press.

Fukuyama, Francis. 1995. *Trust: The Social Virtues and the Creation of Prosperity.* New York: Free Press.

Gardner, Howard. 1993. *Creating Minds: An Anatomy of Creativity Seen Through the Lives of Freud, Einstein, Picasso, Stravinsky, Eliot, Graham, and Gandhi.* New York: Basic Books.

Gelb, Michael J. 1998. *How to Think Like Leonardo da Vinci: Seven Steps to Genius Every Day.* New York: Dell.

Habermas, Jürgen. 1989. *The Structural Transformation of the Public Sphere.* Trans. Thomas Burger. Cambridge, Mass.: MIT Press.

Hansell, Saul. 2001. Web sales of Airline Tickets Are Making Hefty Advances. *New York Times* (July 4).

Hauser, Arnold. 1966. The Social Status of the Renaissance Artist. In *Leonardo da Vinci: Aspects of the Renaissance Genius,* ed. Morris Philipson. New York: Braziller.

Hazemi, Reza, Stephen Hailes, and Steve Wilbur, eds. 1998. *The Digital University: Reinventing the Academy.* London: Springer-Verlag.

Hiltz, Starr Roxanne, and Murray Turoff. 1978. *The Network Nation: Human Communication via Computer.* Reading, Mass.: Addison-Wesley. Rev. ed. Cambridge, Mass.: MIT Press, 1993.

Hughes, John, Val King, Tom Rodden, and Hans Anderson. 1995. The role of ethnography in interactive systems design. *ACM Interactions* 2 (2): 56–65.

Jacoby, Barbara, and Associates. 1996. *Service-Learning in Higher Education.* San Francisco: Jossey-Bass.

Karat, Clare-Marie. 1994. A Business Case Approach to Usability Cost Justification. In *Cost-Justifying Usability,* ed. Randolph Bias and Deborah Mayhew, 45–70. New York: Academic Press.

Kemp, Martin. 2000. *Visualizations: The "Nature" Book of Art and Science.* Berkeley: University of California Press.

Kling, Rob. 1980. Social Analyses of Computing: Theoretical Perspectives in Recent Empirical Research. *ACM Computing Surveys* 12 (March): 61–110.

Kohn, Alfie. 1986. *No Contest: The Case Against Competition.* Boston: Houghton Mifflin.

Kollock, Peter. 1999. The Production of Trust in Online Markets. In *Advances in Group Processes,* Vol. 16, ed. E. J. Lawler, M. Macy, S. Thyne, and H. A. Walker. Greenwich, Conn.: JAI Press.

Kraut, R., W. Scherlis, T. Mukhopadhyay, J. Manning, and S. Kiesler. 1996. The HomeNet Field Trial of Residential Internet Services, *Communications of the ACM* 39 (December): 55–63.

Kuhn, Thomas S. 1962. *The Structure of Scientific Revolutions.* 3d ed. Chicago: University of Chicago Press, 1996.

Kurzweil, Ray. 1999. *The Age of Spiritual Machines.* New York: Viking.

Landauer, Thomas K. 1995. *The Trouble with Computers: Usefulness, Usability, and Productivity.* Cambridge, Mass.: MIT Press.

Laurel, Brenda. 2001. *Utopian Entrepreneur.* Cambridge, Mass.: MIT Press.

Lee, Dick. 2000. *The Customer Relationship Management Survival Guide.* St. Paul, Minn.: High-Yield Marketing.

Leonard, George B. 1968. *Education and Ecstasy.* New York: Dell.

Lessig, Lawrence. 1999. *Code and Other Laws of Cyberspace.* New York: Basic Books.

Leveson, Nancy, and Clark S. Turner. 1993. An Investigation of the Therac–25 Accidents. *IEEE Computer* 26 (7): 18–41.

Levine, Peter. 2000. The Internet and Civil Society. *Reports from the Institute for Philosophy & Public Policy* 20 (4). <http://www.puaf.umd.edu/IPPP/reports/vol20fall00/vol20.html>.

Licklider, J. C. R. 1960. Man-Computer Symbiosis. *IEEE Transactions on Human Factors in Electronics* HFE–1 (March): 4–11.

Marchionini, Gary. 1995. *Information Seeking in Electronic Environments.* New York: Cambridge University Press.

Marchionini, Gary, Maryle Ashley, Lois Korzendorfer. 1993. ACCESS at the Library of Congress. In *Sparks of Innovation in Human-Computer Interaction,* ed. Ben Shneiderman. Norwood, NJ: Ablex Publishers. Available from Intellect Books.

Martin, C. Dianne. 1993. The Myth of the Awesome Thinking Machine. *Communications of the ACM* 36 (4): 120–133.

Maslow, Abraham. 1968. *Toward a Psychology of Being.* 2d ed. New York: Van Nostrand Reinhold.

McConnell, Jeffrey J. 1996. Active Learning and Its Use in Computer Science. *ACM SIGCSE Bulletin* 28, Special Issue, 52–54.

Mehlenbacher, Brad. 1999. Personal Communication.

Millis, Barbara J. 1990. Helping Faculty Build Learning Communities Through Cooperative Groups. In *To Improve the Academy: Resources for Student, Faculty and Institutional Development,* Vol. 10, ed. L. Hilsen, 43–58. Stillwater, Okla.: New Forums Press.

Morino, Mario. 2001. From Access to Outcomes: Raising the Aspirations for Technology Initiatives in Low-Income Communities. <http://www.morino.org>.

Mumford, Lewis. 1934. *Technics and Civilization.* New York: Harcourt Brace.

Nader, Ralph. 1965. *Unsafe at Any Speed: The Designed-in Dangers of the American Automobile.* New York: Grossman.

Naisbitt, John. 1982. *Megatrends: Ten New Directions Transforming Our Lives.* New York: Warner Books.

NAS/NRC (National Academy of Sciences/National Research Council). 1996. *National Science Education Standards.* Washington, D.C.: National Academy Press.

Negroponte, Nicholas. 1995. *Being Digital.* New York: Knopf.

NIE (National Institute of Education). 1984. *Involvement in Learning: Realizing the Potential of American Higher Education.* Final Report of the Study Group on the Conditions for Excellence in American Higher Education. Washington, D.C.

Nielsen, Jakob. 1993. *Usability Engineering.* Boston: Academic Press.

Norman, Don. 1988. *The Psychology of Everyday Things.* New York: Basic Books.

———. 1998. *The Invisible Computer.* Cambridge, Mass.: MIT Press.

Norman, Kent L. 1997. Teaching in the Switched-On Classroom: An Introduction to Electronic Education and HyperCourseware. <http://lap.umd.edu/SOC/sochome.html>.

NTIA (National Telecommunications and Information Administration). U.S. Dept. of Commerce. 1999. Falling Through the Net: Defining the Digital Divide. <http://www.ntia.doc.gov/ntiahome/digitaldivide/>.

———. 2000. Falling Through the Net: Toward Digital Inclusion. <http://www.ntia.doc.gov/ntiahome/fttn00/contents00.html>.

———. 2001. A Nation Online: How Americans Are Expanding Their Use of the Internet. <http:www.ntia.doc.gov/ntiahome/dn/>.

Nuland, Sherwin B. 2000. *Leonardo da Vinci.* New York: Viking.

O'Baoill, Andrew. 2000. Slashdot and the Public Sphere. *First Monday.* <http://firstmonday.org/>.

Okada, Takeshi, and Herbert A. Simon. 1997. Collaborative Discovery in a Scientific Domain. *Cognitive Science* 21 (2): 109–146.

Olson, Gary M., and Judith S. Olson. 1997. Research on Computer Supported Cooperative Work. In *Handbook of Human-Computer Interaction, Second Edition,* ed. M. G. Helander, T. K. Landauer, and P. V. Prabhu, 1433–1456. Amsterdam: Elsevier.

Papert, Seymour. 1980. *Mindstorms: Children, Computers, and Powerful Ideas.* New York: Basic Books.

PCAST (President's Committee of Advisors on Science and Technology). Panel on Educational Technology. 1997. *Report to the President on the Use of Technology to Strengthen K–12 Education in the United States.* Washington, D.C.

Penzias, Arno. 1989. *Ideas and Information.* New York: Simon and Schuster.

Piaget, Jean. 1964. Cognitive Development in Children: The Piaget Papers. In *Piaget Rediscovered: A Report of the Conference on Cognitive Studies and Curriculum Development,* ed. R. E. Ripple and V. N. Rockcastle, 6–48. Ithaca, N.Y.: Ithaca School of Education, Cornell University.

Plaisant, Catherine, Anne Rose, Brett Milash, Seth Widoff, and Ben Shneiderman. 1996. LifeLines: Visualizing Personal Histories. In *Proceedings of CHI 96,* 221–227, 518. New York: ACM Press.

Polya, G. 1957. *How to Solve It: A New Aspect of Mathematical Method.* 2d ed. Garden City, N.Y.: Doubleday Anchor Books.

Preece, Jenny. 2000. *Online Communities: Designing Usability, Supporting Sociability.* Chichester, U.K.: Wiley.

Preece, Jenny, and Diane M. Krichmar. 2002. Online Communities: Social Interaction and Universal Usability. In *Handbook of Human-Computer Interaction,* ed. J. Jacko and A. Sears. Mahwah, N.J.: Erlbaum.

Putnam, Robert. 2000. *Bowling Alone: The Collapse and Revival of American Community.* New York: Simon and Schuster.

Rice, Ronald E., and James E. Katz, eds. 2001. *The Internet and Health Communication: Experience and Expectations.* Thousand Oaks, Calif.: Sage Publications.

Richter, Jean Paul. 1969. *The Literary Works of Leonardo da Vinci.* 3d ed. London: Phaidon.

Schuler, Doug. 1996. *New Community Networks: Wired for Change.* Reading, Mass.: Addison-Wesley.

Shneiderman, Ben. 1980. *Software Psychology: Human Factors in Computer and Information Systems.* Boston: Little, Brown.

———. 1983. Direct Manipulation: A Step Beyond Programming Languages. *IEEE Computer* 16 (8): 57–69.

———. 1989. My Star Wars Plan: A Strategic Education Initiative. *The Computing Teacher* 16 (7): 5.

———. 1992. Education by Engagement and Construction: A Strategic Education Initiative for the Multimedia Renewal of American Education. In *Sociomedia: Hypermedia, Multimedia and the Social Construction of Knowledge,* ed. E. Barrett, 13–26. Cambridge, Mass.: MIT Press.

————. 1998. *Designing the User Interface: Strategies for Effective Human-Computer Interaction.* 3d ed. Reading, Mass.: Addison-Wesley.

Shneiderman, Ben, M. Alavi, K. Norman, and E. Y. Borkowski. 1995. Windows of Opportunity in Electronic Classrooms. *Communications of the ACM* 38 (11): 19–24.

Shneiderman, Ben, E. Y. Borkowski, M. Alavi, and K. Norman. 1998. Emergent Patterns of Teaching/Learning in Electronic Classrooms. *Educational Technology Research & Development* 46 (4): 23–42.

Shneiderman, Ben, and H. Kang. 2000. Direct Annotation: A Drag-and-Drop Strategy for Labeling Photos. In *Proceedings of the International Conference on Information Visualization 2000,* 88–95. Available from IEEE, Los Alamitos, California.

Shneiderman, Ben, and Anne Rose. 1996. Social Impact Statements: Engaging Public Participation in Information Technology Design. In *Proceedings of CQL 96, ACM SIGCAS, Symposium on Computers and the Quality of Life,* 90–96. Also in *Human Values and the Design of Computer Technology,* ed. B. Friedman, 117–133. New York: Cambridge University Press, 1997.

Shulman, S., S. Zavestoski, D. Schlosberg, D. Courard-Hauri, and D. Richards. 2001. Citizen Agenda-Setting: The Electronic Collection and Synthesis of Public Commentary in the Regulatory Rulemaking Process. In *Proceedings of the National Conference for Digital Government Research.* <http://www.isi.edu/dgrc/dgo2001/papers/session3/shulman.pdf>.

Simon, Herbert A. 1996. *The Sciences of the Artificial.* 3d ed. Cambridge, Mass.: MIT Press.

Slavin, Robert. 1990. *Cooperative Learning: Theory, Research, and Practice.* Englewood Cliffs, N.J.: Prentice-Hall.

Snow, C. P. 1993. *The Two Cultures.* Introduction by Stefan Collini. New York: Cambridge University Press. Original lecture 1959.

Soloway, E., S. Jackson, J. Klein, C. Quintana, J. Reed, J. Spitulnik, S. Stratford, S. Studer, S. Jul, J. Eng, and N. Scala. 1996. Learning Theory in Practice: Case Studies of Learner-Centered Design. In *Proceedings of CHI 96,* 189–196. New York: ACM Press.

Sunstein, Cass. 2001. *Republic.com.* Princeton, N.J.: Princeton University Press.

Swift, Ronald S. 2000. *Accelerating Customer Relationships: Using CRM and Relationship Technologies.* Upper Saddle River, N.J.: Prentice-Hall.

Turner, A. Richard. 1994. *Inventing Leonardo.* Berkeley: University of California Press.

Uslaner, Eric. 2001. *The Moral Foundations of Trust.* New York: Cambridge University Press.

Van Tassel, Joan. 1994. Yakety-Yak, Do Talk Back! *Wired Magazine.* <http://www.wired.com/wired/archive/2.01/pen.html>.

Varley, Pamela. 1991. Electronic Democracy. *Technology Review* 94 (November): 42–51.

Vasari, Giorgio. 1998. *Lives of the Artists.* Trans. Julia C. Bondanella and Peter Bondanella. New York: Oxford University Press.

Vygotsky, L. 1934. *Thought and Language.* Trans. A. Kozulin. Cambridge, Mass.: MIT Press, 1986.

Wallace, Robert. 1966. *The World of Leonardo.* New York: Time-Life Books.

Wees, W. R. 1971. *Nobody Can Teach Anybody Anything.* New York: Doubleday.

Weinberg, Gerald M. 1971. *The Psychology of Computer Programming*. New York: Van Nostrand Reinhold.

Weizenbaum, Joseph. 1976. *Computer Power and Human Reason: From Judgment to Calculation*. San Francisco: W. H. Freeman.

White, Michael. 2000. *Leonardo: The First Scientist*. New York: St. Martin's Press.

Wildmoon, K. C. 1997. Peace Through E-mail: Wired Activists Find Strength in Cyberspace. CNN Interactive. <http://www.cnn.com/specials/1997/nobel.prize/ stories/internet.coalition/ index.html>.

Wilhelm, Anthony. 2000. *Democracy in the Digital Age: Challenges to Political Life in Cyberspace*. New York: Routledge.

Winograd, Terry, and Fernando Flores. 1986. *Understanding Computers and Cognition: A New Foundation for Design*. Norwood, N.J.: Ablex.

Wright, Frank Lloyd. 1953. *The Future of Architecture*. New York: Horizon Press.

Zuboff, Shoshanna. 1988. *In the Age of the Smart Machine: The Future of Work and Power*. New York: Basic Books.